Phylogenetic Systematics

UNIVERSITY OF ILLINOIS PRESS, URBANA · CHICAGO · LONDON, 1966

PHYLO-GENETIC SYSTEMATICS

By Willi Hennig

translated by D. Dwight Davis and Rainer Zangerl

Illinois Reissue, 1999

Library of Congress Cataloging-in-Publication Data
Hennig, Willi, 1913–
Phylogenetic systematics.
Bibliography: p.
1. Zoology—Classification. 2. Phylogeny
I. Title.
OL351.H413   1979   591'.01'2   78-31969
ISBN 0-252-06814-9 / 978-0-252-06814-0

Preface

This book is not a translation of Hennig's *Grundzüge einer Theorie der phylogenetischen Systematik* (1950). That work has been extensively revised by the author and much of it has been completely rewritten. The latter is the manuscript that has been translated. Consequently the text appears here for the first time; it has not been published in German.

My own involvement in this project was the result of my ability to master the language, but it was my close friend and colleague, the late Dr. D. Dwight Davis, who had assumed the responsibility for translating the work. Following completion of the rough translation, Davis spent a great deal of time editing the text, but unhappily could not finish the task. The responsibility for the final form of the translation now rests with me.

The translation of a work such as this poses a great many linguistic difficulties in the sense that for some German concepts there simply are no English equivalents. One of the worst examples of this is the concept of *Gesetzmässigkeit*, widely used in the German technical literature. We have used the literal "conformity to law," fully realizing that there is no perfect congruence between these concepts. For example, a Gesetzmässigkeit may be no more than the repetitive occurrence of a specific phenomenon.

RAINER ZANGERL

Foreword

Phylogenetic Systematics, first published in 1966, marks a turning point in the history of Systematic Biology. It appeared in North America during a time of confusion and debate over establishments generally, and particularly over the principles of this, the most basic and seemingly well established of biological disciplines. It was one of a dozen treatises published during a fifteen-year period of uncertainty about methods and goals (Simpson, 1961; Davis and Haywood, 1963; Sokal and Sneath, 1963; Croizat, 1964; Traub, 1964; Blackwelder, 1967; Mayr, 1969; Crowson, 1970; Jardine and Sibson, 1971; Boyden, 1973; Sneath and Sokal, 1973; Ross, 1974; Clifford and Stephenson, 1975). The fretful and searching mood of the times was also reflected in scientific journals, such as *Systematic Zoology* and *Taxon,* with their focus on methods and goals, and in 1976 by the birth of a new journal with the same focus, *Systematic Botany.*

Books, like people born in a context of contention, sometimes become notorious through the friends they make and the enemies they antagonize. *Phylogenetic Systematics* has had its share of each. An early version of the book was published in German (Hennig, 1950). Over the next fifteen years it served as the organizing force for the research of a Swedish entomologist (Brundin, 1966) whose biogeographic results, as noted by one early reviewer (Byers, 1969), coupled Hennig's view of systematics with the then-developing revolution in geology known as plate tectonics. The initial response to *Phylogenetic Systematics,* as manifested by many other reviewers and commentators, generally failed to perceive the possibilities in applying Hennig's approach—even though the possibilities were already and amply realized by Brundin (Cain, 1967; Horowitz, 1967; Sokal, 1967; Bock, 1968; Peters, 1968; Colless, 1969; Darlington, 1970; Michener, 1970; Ashlock, 1974; Mayr,

1974; Simpson, 1975; Blackwelder, 1977; Banarescu, 1978; Eichler, 1978; Van Valen, 1978). The problem with those reviewers and commentators, of course, is that they were firm adherents of, and some of them had vested interests in, other points of view.

In modern science the useful life of any technical and non-popular publication is apt to be short. A dozen years normally suffice for birth, maturation, senescence, death, and forgetful oblivion. Yet in the years since 1966, *Phylogenetic Systematics* has not been superseded. Rather, it has slowly worked its way, despite its critics, to the top of the heaped publications of the period. And unless we are mistaken, its role is not yet played out. It could become the basic document of its half-century—an assertion justified by the many dog-eared and much-annotated copies that survive from the original edition.

The continuing relevance of *Phylogenetic Systematics* is closely tied to the relevance of its subject—a certain view of the methods and goals of Systematic Biology that has since been termed "cladistics." Its relevance is currently being explored in seven general areas: 1) philosophy of science, with particular reference to the views of Karl Popper (Wiley, 1975); 2) historical background of cladistics, which has some similarity with the theoretical writings of the early twentieth-century Italian zoologist Daniele Rosa (Croizat, 1976; Baroni-Urbani, 1977), and which echoes Darwin's thoughts on the potential impact of evolutionary theory on systematics (Nelson, 1974); 3) relations between cladistics and other historical disciplines (Platnick and Cameron, 1977); 4) implications within Systematic Biology (Bonde, 1977; Nelson, 1979); 5) biogeographical synthesis, with particular reference to the views of Leon Croizat (Rosen, 1978); 6) computerized, or "numerical," cladistics (Farris, Kluge, and Eckardt, 1970; Farris, 1977); 7) molecular cladistics, based on amino-acid and nucleotide sequences—for which the method of cladistics was independently developed. but which is integrable with Hennig's approach (Goodman, 1976; Fitch, 1977). Exploration of these seven areas has only recently begun, and is by no means exhausted. Yet biogeographical synthesis, itself with a historical background of interesting and surprising complexity when viewed in the cladistic aspect (Nelson, 1978), is already emerging as the probable focus of Systematic Biology during the remaining decades of the twentieth century.

The relevance of cladistics in particular areas is manifested by the increasing number of revisionary studies that include "cladistic analysis"—either in their title or in their substance. The time has, perhaps, arrived when cladistics has become the standard approach toward systematics and taxonomy of animals if not yet of plants. It remains to be seen whether cladistics will have a similar impact upon systematic botany, but there are indications that it will (Bremer and Wanntorp, 1978; Parenti, 1979).

When Thompson (1958), in the introduction to the Everyman Edition of Darwin's *Origin of Species*, wrote that the "success of Darwinism was accompanied by a decline in scientific integrity," he was decrying not only the absence of constraints in the manufacture of evolutionary theories, but also the absence of some standard against which evolutionary explanation could be measured. One of Hennig's primary goals was to provide such a standard in the form of statements about the relationships observable in the living world. To Hennig, the relationships are expressed in a hierarchy of organisms and the homologous parts of organisms—a natural order of sets and subsets. For Hennig, discovery of natural order, as expressed in a diagram, sharply defines these relationships, and therefore constrains the number and kinds of possible explanations of them. He intended evolutionary explanation to depend not merely on the limits of an investigator's imaginativeness but also on what the evidence would allow: nature would instruct the investigator, not the other way around.

Hennig established a criterion of demarcation between science and metaphysics at a time when neo-Darwinism had attained a sort of metaphysical pinnacle by imposing a burden of subjectivity and tautology on nature's observable hierarchy. Encumbered with vague and slippery ideas about adaptation, fitness, biological species, and natural selection, neo-Darwinism (summed up in the "evolutionary" systematics of Mayr and Simpson) not only lacked a definable investigatory method, but came to depend, both for evolutionary interpretation and classification, on consensus or authority. Several authors besides Hennig recognized the shortcomings of neo-Darwinism and attempted to establish similar or entirely different criteria of demarcation (e.g., Crowson, Sokal and Sneath, Blackwelder), but for us Hennig was the most successful because his method is simple, explicit, and tied to nature's hierarchy.

Emphasis on the relational aspects of nature, as expressed in a branching diagram (cladogram) of its hierarchical structure, had a profound influence on the conduct and limits of systematic studies. It became clear that the hierarchy itself does not necessarily contain evidence that evolution has taken place. If the hierarchy is *assumed* to be the result of an evolutionary process, criteria of a genealogical nexus are provided by each addition of empirical data. The idea of organic evolution was thereby changed from a narrative about the history of life to a scientific theory about how nature came to be ordered in its particular hierarchical structure. It also became clear that fossils, long held in some sense to be proof that evolution had occurred, were merely parts of this hierarchy, subject to the same constraints of interpretation as living organisms (Patterson and Rosen, 1977). This realization marked the beginning of the end of traditional paleontology and its search for ancestors. Paleontology becomes, in that realization, no longer the keeper of profound evolutionary truths but rather the keeper of extinct parts of nature's hierarchy.

If paleontology has lost one claim to our attention, the modern principles of cladistics that emerged from Hennig's exposition give it a grander, if more austere, claim. As seen by Hennig, the fossil record, rather than forming the basis for independent hierarchies that dictate the interpretation of extant species, is subsidiary to a comprehensive hierarchy of extant species within which the record provides a time dimension. The importance of fossils aligned among their extant relatives becomes evident in historical interpretation of the hierarchy in time and space. In the distributional history of a group of organisms, the furcations in a branching diagram may be considered as speciation events made possible by geographic change. The relevance of one or another series of paleogeographic events, which are themselves amenable to cladistic analysis, is tested, in part, by congruence between the ages of fossils within the biological hierarchy and the ages of the parts of the paleogeographic sequences that might have influenced the course of speciation (Platnick and Nelson, 1978).

It is clear, at least to us, that cladistics is *the* general method of historical science. Cladistics is not an exotic and idiosyncratic method as claimed by some self-styled neo-Darwinians. The broad application of cladistic principles to a truly synthetic natural science integrating neontology, paleontology, biogeography, and geology might have been long delayed without the stimulus provided by *Phylogenetic Systematics*.

The value of cladistics in achieving unity in method and purpose between paleontology and neontology was nowhere better demonstrated than in a Linnean Society of London symposium held in June, 1972. In the preface to the published proceedings of the symposium, the editors noted: "Readers of the volume may be struck by similarities in the way that most of the contributors have approached the problem of investigating and describing relationships. These methods, more precise and explicit that those traditionally used, are due to the influence of Hennig, not to editorial pressure" (Greenwood, Miles, and Patterson, 1973).

This reprinting of *Phylogenetic Systematics* has been initiated through the action of Fellows of the Linnean Society of London—a venerable organization whose members have somehow managed repeatedly to penetrate the mists of their own traditions, and to maintain a sense of current and forthcoming relevance in the science to which that society is dedicated. The Linnean Society is one of two institutions that have acknowledged the unique value of *Phylogenetic Systematics* by presenting its author with their highest award: the *Linnean Gold Medal* (Anon., 1974) and the *Gold Medal for Distinguished Achievement in Science* of the American Museum of Natural History (Nicholson, 1975).

Willi Hennig did not live to accrue further honors; he died in 1976, at the age of sixty-three. After receiving his doctorate in Leipzig, he worked in the Deutsches Entomologisches Institut in Berlin, specializing in Diptera. There he produced his monumental *Larvenformen der Dipteren* (1948-52), the 1950 book that foreshadowed this one, and his work on the Diptera of New Zealand (1960), in which he first showed the relevance of cladistics to biogeography. After fourteen years as head of the Department of Systematic Entomology at the Entomologisches Institut, Hennig moved in 1963 to the Staatliches Museum in Stuttgart, as director for phylogenetic research. There he wrote *Die Stammesgeschichte der Insekten* (1969), the full impact of which is yet to be felt in entomology. An English translation of that book is shortly to be published, and will undoubtedly cause renewed interest in *Phylogenetic Systematics*.

As we see it, the significance of this reprinting is that the possibilities opened by Hennig's approach to systematics are still being explored, and the full consequences of the application of his concepts in evolutionary biology are yet to be discovered.

Department of Ichthyology   DONN E. ROSEN
American Museum of Natural History   GARETH NELSON
October, 1978   COLIN PATTERSON

References

Anon. 1974. The Linnean gold medals. *Biological Journal of the Linnean Society* (London) 6, no. 4:375-77.

Ashlock, P. D. 1974. The uses of cladistics. *Annual Review of Ecology and Systematics* 5:81-99.

Banarescu, P. 1978. Some critical reflexions on Hennig's phyletical concepts. *Zeitschrift für zoologische Systematik und Evolutionsforschung* 16, no. 2:91-101.

Baroni-Urbani, C. 1977. Hologenesis, phylogenetic systematics, and evolution. *Systematic Zoology* 26, no. 3:343-46.

Blackwelder, R. E. 1967. *Taxonomy.* John Wiley and Sons, Inc., New York.

———. 1977. Twenty-five years of taxonomy. *Systematic Zoology* 26, no. 2: 107-37.

Bock, W. J. 1968. Phylogenetic systematics, cladistics, and evolution. *Evolution* 22, no. 3:646-48.

Bonde, N. 1977. Cladistic classification as applied to vertebrates. In M. K. Hecht, P. C. Goody, and B. M. Hecht, eds., *Major patterns in vertebrate evolution* (Plenum Press, New York), pp. 741-804.

Boyden, A. 1973. *Perspectives in zoology.* Pergamon Press, Oxford.

Bremer, K., and H.-E. Wanntorp. 1978. Phylogenetic systematics in botany. *Taxon* (in press).

Brundin, L. 1966. Transantarctic relationships and their significance as evidenced by chironomid midges. *Kungliga Svenska Vetenskapsakademiens Handlingar,* 4th ser. 11, no. 1:1-472.

Byers, G. W. 1969. [Book reviews]: *Systematic Zoology* 18, no. 1:105-7.

Cain, A. J. 1967. One phylogenetic system. *Nature* (London), 216, no. 5113:412-13.

Clifford, H. T., and W. Stephenson. 1975. *An introduction to numerical classification.* Academic Press, New York.

Colless, D. H. 1969. The interpretation of Hennig's "phylogenetic systematics." *Systematic Zoology* 18, no. 1:134-44.

Croizat, L. 1964. *Space, time, form: the biological synthesis.* Published by the author, Caracas.

―――. 1976. Biogeografía analítica y sintética ("panbiogeografía") de las Américas. *Biblioteca de Ciencias Físicas, Mathemáticas y Naturales* (Caracas) 15-16:1-454, 455-890.

Crowson, R. A. 1970. *Classification and biology.* Atherton Press, New York.

Darlington, P. J., Jr. 1970. A practical criticism of Hennig-Brundin "phylogenetic systematics" and Antarctic biogeography. *Systematic Zoology* 19, no. 1:1-18.

Davis, P. H., and V. H. Haywood. 1963. *Principles of angiosperm taxonomy.* Oliver and Boyd, Edinburgh.

Eichler, W. 1978. Kritische Einwände gegen die Hennigsche kladistische Systematik. *Biologische Rundschau* 16, no. 3:175-85.

Farris, J. S. 1977. On the phenetic approach to vertebrate classification. In M. K. Hecht, P. C. Goody, and B. M. Hecht, eds., *Major patterns in vertebrate evolution* (Plenum Press, New York), pp. 823-50.

―――, A. G. Kluge, and M. J. Eckardt. 1970. A numerical approach to phylogenetic systematics. *Systematic Zoology* 19, no. 2: 172-89.

Fitch, W. M. 1977. The phyletic interpretation of macromolecular sequence information: simple methods. In M. K. Hecht, P. C. Goody, and B. M. Hecht, eds., *Major patterns in vertebrate evolution* (Plenum Press, New York), pp. 169-204.

Goodman, M. 1976. Towards a genealogical description of the Primates. In M. Goodman, R. E. Tashian, and J. H. Tashian, *Molecular anthropology* (Plenum Press, New York), pp. 321-53.

Greenwood, P. H., R. S. Miles, and C. Patterson. 1973. Preface to *Interrelationships of fishes. Supplement no. 1 to the Zoological Journal of the Linnean Society* 53.

Hennig, W. 1950. *Grundzüge einer Theorie der phylogenetischen Systematik.* Deutscher Zentralverlag, Berlin.

Horowitz, A. S. 1967. [Book review]: *Journal of Paleontology* 41, no. 6:1569.

Jardine, N., and R. Sibson. 1971. *Mathematical taxonomy.* John Wiley and Sons, Inc., New York.

Mayr, E. 1969. *Principles of systematic zoology.* McGraw-Hill Book Co., New York.

―――. 1974. Cladistic analysis or cladistic classification? *Zeitschrift für zoologische Systematik und Evolutionsforschung* 12, no. 2:94-128.

Michener, C. D. 1970. Diverse approaches to systematics. *Evolutionary Biology* 4:1-38.

Nelson, G. 1974. Darwin-Hennig classification. *Systematic Zoology* 23, no. 3: 452-58.

―――. 1978. From Candolle to Croizat: comments on the history of biogeography. *Journal of the History of Biology* 11, no. 2:269-305.

―――. 1979. Cladistic analysis and synthesis: principles and definitions. *Systematic Zoology* 28, no. 1:1-21.

Nicholson, T. D. 1975. Report of the director. *Annual Report of the American Museum of Natural History*, no. 106:6-7.

Parenti, L. 1979. Cladogramming the land plants. *Systematic Botany* (submitted).

Patterson, C., and D. E. Rosen. 1977. Review of ichthyodectiform and other Mesozoic teleost fishes and the theory and practice of classifying fossils. *Bulletin of the American Museum of Natural History* 158, art. 2:81-172.

Peters, J. A. 1968. Phylogenetic systematics. *Copeia*, no. 1:199-200.

Platnick, N. I., and H. D. Cameron. 1977. Cladistic methods in textual, linguistic, and phylogenetic analysis. *Systematic Zoology* 26, no. 4:380-85.

―――, and G. Nelson. 1978. A method of analysis for historical biogeography. *Systematic Zoology* 27, no. 1:1-16.

Rosen, D. E. 1978. Vicariant patterns and historical explanation in biogeography. *Systematic Zoology* 27, no. 2:159-88.

Ross, H. H. 1974. *Biological systematics.* Addison-Wesley Publishing Co., Inc., Reading, Mass.

Simpson, G. G. 1961. *Principles of animal taxonomy.* Columbia University Press, New York.

―――. 1975. Recent advances in methods of phylogenetic inference. In W. P. Luckett and F. Szalay, eds., *Phylogeny of the Primates* (Plenum Press, New York), pp. 3-19.

Sneath, P. H. A., and R. R. Sokal. 1973. *Numerical taxonomy.* W. H. Freeman and Co., San Francisco.

Sokal, R. R. 1967. Principles of taxonomy. *Science* (Washington, D.C.) 156, no. 3780:1356.

―――, and P. H. A. Sneath. 1963. *Principles of numerical taxonomy.* W. H. Freeman and Co., San Francisco.

Thompson, W. R. 1958. Introduction to *The Origin of Species,* 6th ed. Everyman's Library, E. P. Dutton, New York.

Traub, H. P. 1964. *Lineagics.* The American Plant Life Society, La Jolla.

Van Valen, L. 1978. Why not be a cladist. *Evolutionary Theory* 3, no. 5:285-99.

Wiley, E. O. 1975. Karl R. Popper, systematics, and classification. *Systematic Zoology* 24, no. 2:233-43.

Contents

I

The position of systematics among the biological sciences

GENERAL CONCEPT OF SYSTEMATICS

In no other science are the contrasts and struggle for survival among subdivisions of the field so strong as in biology. This is at least partly because the problems, and therefore the methods, are more varied in biology than in any other science. This diversity demands intense concentration and specialization by anyone dealing with any of its numerous problems, which in turn leads to considerable isolation of its subdivisions. Rivalries among biological disciplines seem less sharp now than they once were, at least as far as relative values are concerned, but they are still important because they determine the recognition and representation of the several disciplines in colleges and universities, and thus in the final analysis the economic basis for their continuation.

If in this struggle for survival biological systematics has recently lost ground to other and, as is often heard, younger and more modern disciplines, this is not so much because of the limited practical or theoretical importance of systematics as because systematists have not correctly understood how to present its importance in the general field of biology, and to establish a unified system of instruction in its problems, tasks, and methods.

This deficiency and the resulting disadvantages have been correctly diagnosed from various quarters. Thus von Wettstein (1937) wrote: "In the past two decades systematics has fallen back to the rear lines in comparison with other fields of biological research, especially genetics and physiology. The reason for this is not so much lesser importance of systematics or the exhaustion of its problems, as many believe . . . but rather because the research methods hitherto used in systematics are no longer adequate to meet the greatly increased requirements resulting from knowledge in genetics and physiology. Except for a few representatives of the old school, the search for new methods and directions is observable everywhere." Paramonow (1935) said: "The shortcomings we have mentioned are reinforced by the completely unsatisfactory situation with respect to the theory of systematics. The theory is very little worked over; there is no book that deals with the principles of systematics in adequate detail." Martini (1938) also considered it "probable that the time is ripe for a discussion of the foundations and methods of biological systematics, not as an end in itself but from the viewpoint of a member science of biology as a whole, and to give new vigor to the wonderful science of systematics."

In the twenty years since these statements were written a relatively large number of works dealing with the theory of biological systematics have appeared. They are cited and critically evaluated in the reports of Günther (1956, 1962). Yet we are still far from having a basic, comprehensive, and generally acknowledged theory of biological systematics. This is primarily because it is almost impossible to summarize the details of all areas of this widely ramified science and be equally competent in all. Consequently, it is not to be expected that a comprehensive theory of systematics can be created at one stroke.

The peculiar place and significance of systematics in biology is clearly recognizable only if an account of the place of biology itself among the natural sciences is first presented.

As Max Hartmann explains in his *Allgemeine Biologie* (1947), the natural sciences are usually divided into the exact or explanatory sciences and the descriptive sciences. Windelband and Rickert believed that this distinction between the nomothetic and ideographic sciences expresses the difference between the natural sciences on the one hand, and the cultural and mental sciences on the other. But, according to Hartmann, among the natural sciences only physics is a "pure science of laws," to be regarded as the "elementary general science of the world of phenomena," which must be "contrasted with all other disciplines of natural science as the basic discipline" because all other disciplines, collectively or individually, have at least some ideographic aspects.

Hartmann sees the special position of physics as based on the fact that it deals with natural processes, and not with individual natural bodies existing in space and time. The task of all other natural sciences, including biology, is "to present a complete and exhaustive account of a single more or less extensive event in temporally limited actuality." Hartmann adds, as a decisive characteristic of all natural sciences except physics, that they have to do not only with processes, but above all with "individualized bodies," with "relatively stable corporal systems." We may add that this is true of biology, whose special objects of research are living natural bodies.

Up to this point we can agree with Hartmann. But it is very questionable whether "systematics," "description," "ordering," and "cataloging" can be contrasted with "rationalization of the world of phenomena," and equated with "leading back to physical laws," as Hartmann does. In philosophy, "systematics" is generally used as synonymous with "nomothetic," "theoretical," "lawful," etc., as opposed to "ideographic," "historical," "descriptive," etc. (see Ziehen, *Erkenntnistheorie*, e.g., 1934, p. 178, etc.). Things, phenomena, and processes are here regarded as "systematic" that are not to be understood as the simple description of accessible isolated phenomena, but as elements of an ordering of things, phenomena, or processes according to law. Consequently "science" has been correctly defined as "the systematic orientation of man in his environment."

Moreover, pairs of concepts such as descriptive and explanatory, ideographic and nomothetic, seem to us completely unusable in sorting the natural sciences. "Each explanation demands in itself a new explanation, i.e., the search for still more inclusive relationships, compared to which they look like mere 'description,' like the recognition of a simple circumstance of fact" (von Bertalanffy, 1932). Thus description and explanation do not divide the different sciences from each other, but in each science and in each partial discipline they are interwoven in an inseparable way. This also applies, as we will demonstrate in detail, to the biological discipline that is usually called systematics. If we arbitrarily designate only physical laws as "laws" in natural science, then we can agree with Hartmann that biology is a science of laws, a "nomothetic science, insofar as it tries to analyze the stationary processes in their sharply individualized forms in terms of physical regularities, and seeks to analyze causally the broad range of the processes of change in form." But we cannot continue with Hartmann, and say that biology is a "systematic science of ordering, in that it seeks to include the tremendous variety of individual organic forms in an ordered system," without adding that "order" and "systematics" in this sense are not equivalent to description, but also include explanation and rationalization.

THE SPECIAL TASKS OF BIOLOGICAL SYSTEMATICS

In the preceding section it was shown that systematics in the most general sense of the word is equivalent to order, rationalization, and in a certain context to explanation of the world of phenomena; and that in this sense systematics is a very broad task of all natural sciences, and particularly of all biological disciplines. It now remains to determine which part of the general task of ordering falls to the discipline that is generally simply called systematics.

Ziehen (1939) defines "order" as "the totality of progressively graduated vicinal similarities of more or less determined positional relationships of several or many, even an infinite number, maximally all, 'somethings' within a finite or infinite whole." As "position" he understands "a more or less definite relation of a simple or complex something to other somethings that belong to a unified whole in regard to quality, intensity, locality, temporality, or number."

In this definition it seems to me particularly essential that the task of "ordering" (and what means the same thing, of systematics), lies in considering the unit as a member of an ordered whole. It is a fact, particularly sharply worked out by O. Spann, that no unit exists as a member of only one whole.

In terms of quantum theory, or class theory, each single element may occur in differently composed partial quantities, or partial classes, within a whole quantity, depending on which of the many relationships existing among the elements of the total quantity are chosen to form the partial quantities.

This also applies to the objects of biological systematics; the "animate natural things" in their totality form a "multidimensional multiplicity" if, with Ziehen (1939), we denote the major directions of their diversity (thus, so to speak, the main directions of their varying position within the "realm of their peculiarity") as "dimensions." One can just as well say that the "difference in position" of the animated natural things is brought about by the differences in the relationships between them.

Therefore it is possible to arrange animated natural things in numerous different systems, depending on which of these different relationships has been investigated. The differences among all these systems are determined by the particular relationships of which they are a concrete expression. All these different systems are, fundamentally, equally justified so long as they are a proper expression of the membership position that an object of nature possesses within the framework of the totality, for the dimension that was chosen as the basis for the particular system. On the other hand, every system is fully and unequivocally determined by the kind of relations within the multidimensional multiplicity of which it is an expression; or to express it another way, by the structure of that particular totality of which the individual things should be understood to be members, so that there cannot be any arbitrariness in the formulation of the system.

The different systems in which the member positions of concrete identical individual things are expressed within the framework of different "totalities" are not unrelated to one another. The relations between them, whose nature is determined by the relationship in which the different dimensions of the multidimensional complexity of the individual things exist to one another, can themselves be made the subject of scientific systematic investigation. On the other hand, it is not basically a scientific task to combine several systems so created, because one and the same object cannot be presented and understood at the same time in its position as a member of different totalities. At the blueprint stage, plan and elevation of a building are of equal value and equally justified, and there is a definite relationship between them. But they cannot be combined in a simple presentation, at least not without losing the value that each has for itself.

If we now try to apply these general ideas to the concrete circumstances of biology, we find that the task of biological systematics would be simplest if there were only a single temporally unchangeable organism. But even in such a case systematics would be necessary according to what was said above. Such systematics could be, and would have to be, restricted to investigating the position of the different characteristics of this organism and the various processes that take their course within the framework of the total characters of life. But since there are an enormous number of different organisms that are constantly changing in time, the task of biological systematics is vastly more complicated. Each organism may be conceived as a member of the totality of all organisms in a great variety of ways, depending on whether this totality is investigated as a living community, as a community of descent, as the bearer of the physiological characters of life, as a chorologically differentiated unit, or in still other ways. The classification of organisms or specific groups of organisms as parasites, saprophytes, blood suckers, predators, carnivores, phytophages, etc.; into lung-, trachea-, or gill-breathers, etc.; into diggers of the digging wasp type, mole type, and earthworm type; into homoiothermous or poikilothermous; into inhabitants of the palearctic, neotropical, and ethiopian regions, etc., are partial pieces of such systematic presentations that have been carried out for different dimensions of the multidimensional multiplicity. It does not matter that these "systems" are not usually carried out in extreme detail, but usually are satisfied with relatively coarse groupings. Basically it seems clear that what we are dealing with are elements of the different possible and necessary systems in biology. In many biological disciplines it would be advantageous if the systematic presentation of their underlying relations were less superficial and carried out in a more systematic manner than is usually done.

The different systems in biology are not unrelated, because in the final analysis they contain the same objects as the ultimate elements (for example, the mole under the insectivorous animals, the lung-breathers, the diggers of the mole type, the homoiotherms, and the inhabitants of the palearctic region). The differences between the systems develop only because the basic objects are the bearers of different characteristics of morphological, ecological, physiological, or other nature, and thus appear as members of a physiological system (or of different physiological systems, depending on whether one takes into account all physiological characters or only a few of them as the totality), of one or several ecological, morphological, and other systems.

The opinion exists, as we shall show, that the individual organism, if not even the species, must be thought of as the ultimate and most basic element of biological systematics. This is particularly true of morphological or phylogenetic systematics. Thus this element (individual or species) would have to recur in an identical manner in all kinds of possible systems. But it is one of the oldest and simplest insights of biology that the individuals with all their characters and peculiarities are not constant units but change in many ways in the course of even short periods of time. Recently this fact has again been emphasized particularly strongly, and has been placed in the foreground of interest as an extremely dynamic interpretation not only of the processes of life but also that of the living form: "each describable single form is only an arbitrary portion of the whole that is determined by the point in time chosen" (Torrey, from a review by Dabelow). Although the insight, often strongly emphasized, that the distinction between structure and process has only a conventional and anthropocentric character (von Bertalanffy), Torrey (1939) is correct when he complains that the biologist still works far too little with those concepts, familiar to the physicist and mathematician, of the four-dimensional continuum of space and time.

The significance for biological systematics that attaches to the variability of the individual in time is that, strictly speaking, one and the same individual assumes a different place in most systems at different times of its life. At first sight this fact seems peculiar, but it immediately becomes evident if it is clarified by a very simple example. The larva of the May beetle assumes an entirely different place in an ecological system, that is, in a system that seeks to present the whole of all living organisms as a community, than the sexually mature beetle does. In this system the larva would be more closely associated with other animals that live in the ground and eat roots than with the imago of the May beetle into which it later develops. The imago would be more closely associated with other (flying and leaf-eating) animals. The same applies in countless similar cases for most of the imaginable morphological and physiological systems. From these ideas it follows that we should not regard the organism or the individual (not to speak of the species) as the ultimate element of the biological system. Rather it should be the organism or the individual at a particular point of time, or even better, during a certain, theoretically infinitely small, period of its life. We will call this element of all biological systematics, for the sake of brevity, the *character-bearing semaphoront*. Definition of a semaphoront as the individual during a certain, however brief, period of time (not "at a point in time") has the advantage that it may be thought of more simply as acting and showing evidence of life processes. No generally applicable statements can be made about how long a semaphoront exists as a constant systematically useful entity. It depends on the rate at which its different characters change. In the maximum extreme it would be approximately congruent with the duration of the life of the individual. In many other cases, particularly in organisms that undergo metamorphic and cyclomorphotic processes, it would be notably shorter.

The contention that we should not regard the individual but an even smaller unit, the semaphoront, as the element of biological systematics at first seems rather surprising and perhaps even an artificial construction. It becomes more plausible if one considers that the recognition that the different stages of metamorphosis (larva, pupa, imago) belong to a specific individual life cycle is often the result of complex scientific investigation. Therefore to a science approaching the diversity of living things in a completely unbiased way it is, at least in many cases, actually not the individuals but the semaphoronts that are basic. The concept of the semaphoront, which was coined in order to gain a firm starting point for questions relating to the task of biological systematics, has since been picked up by some ecologists and found useful.

The morphological characters of its spatial, three-dimensional body are not the only properties of a semaphoront. Rather these properties encompass the totality of its physiological, morphological, and psychological (ethological) characters. We will call the totality of all these characters simply the total form (or the holomorphy) of the semaphoront, which thus is to be regarded as a multidimensional construct. Of the many peculiarities of a single semaphoront, only those are of interest for biological systematics that apply to it alone and those that it has in common with a limited number of other semaphoronts, but not those that apply to all existing semaphoronts. We will call those peculiarities that distinguish a semaphoront (or a group of semaphoronts) from other sema- phoronts "characters," keeping in mind that this designation will never include merely morphological characters in the narrow sense, but always means the multidimensional totality.

In the semaphoront we have found the element that must be identical in all conceivable systems of biology insofar as they are related to the living natural objects and not to particular properties of life or to specific processes of life. It is therefore possible to investigate the relations between the different systems, which in themselves are completely and basically equally justified and equally necessary. This is done most usefully by choosing one system as the general reference system with which all the others are compared. Creating such a general reference system, and investigating the relations that extend from it to all other possible and necessary systems in biology, is the task of systematics.

We have characterized the tasks of systematics as the investigation and pres- entation of all relations that exist among living natural objects, and systematics as all scientific activity that aims at ordering and rationalizing the world of phenomena. It may seem that we have fallen into the error of trying to broaden the concept of systematics far beyond the scope usually accorded it in biology. The representatives of many biological disciplines, in comprehensive presenta- tions, often tend to broaden the concept of their own discipline to such an exaggerated extent that it finally seems to include all biology. We will consci- entiously try to avoid this mistake. On the other hand, it was unquestionably necessary to present in more detail the essential and original sense of the con- cept of systematics as it is currently used in many sciences outside biology. In biology even today there are systematists who know nothing better to say of biological systematics than that it is necessary for science in the same way as a card catalog is for a library (Heintz 1939). Even von Bertalanffy (1932) writes that systematics stands at the beginning of biology and that "its goal is to produce as complete and accurate a catalog as possible of plant and animal species." Such a systematics would not be a science. Hertwig (1914) already pointed this out, but without drawing the necessary conclusions, and Zimmermann is entirely correct when he notes that "such a functional grouping is a technical, not a scientific, problem." It is very peculiar that most biologists who have voiced opinions do not seem to understand that a purely "functional grouping" in this sense cannot have any significance for biology as a science, and hence would be entirely lacking in sense and usefulness.

Hertwig (1914) and Plate (1914) stated that systematics "pursues primarily practical aims," and should make it possible to find a species quickly (Plate), and that its task is "to make order in the chaos of organisms that surround us and to provide the means, in a well-differentiated systematic arrangement, to orient oneself within the multiplicity." Unfortunately even today this represents the opinion of many systematists. This is seen, for example, in a recently expressed view by Borgmeier (1955) that goes back to Horn (1929a): "Taxonomy is ordering without saying anything about the way in which this order has come into being."

We must disagree most strongly with this. In reality, nothing at all is achieved by a completely nontheoretical ordering of organisms. All results of biological investigation, in whatever partial discipline they may have been gathered, have meaning only if a realm of applicability can be seen in them that extends beyond the individuals (single organisms) from which they were derived. Thus the results must have validity for certain groups of individuals (semaphoront groups). These groups of individuals may pertain to a physiological (homoiothermy, for example), ecological (parasites), phylogenetic (insects), or any other constructed system. The relationships that led to the proposal of the group for whose members conformity to law is assumed (and which were determined from individual organisms), must always be exactly determined. Thus for both scientific and applied biology the only systems that have any significance are those based on definite natural and exactly determined relationships between semaphoronts. These are, therefore, systems whose structure of ordering is exactly defined. I have tried to show above that scientific biological systematics has to do only with systems of this kind. In the following we will restrict the concept of systematics, in conformity with current use, not to the creation of all conceivable systems that can be based on the relations among organisms, but to the creation of a general reference system and the pursuit of the relationships between this and other possible systems. This does not alter the basic fact that systematics fundamentally means any investigation of relations between natural things and natural processes insofar as they have the character of conformity to law.

The tenacious persistence of the idea that biological systematics should deal primarily with creating an inventory catalog of all animal and plant species is probably attributable to the fact that the species category seems to occupy a particularly preferred position within the realm of applicability of many biological conformities to law. It will be shown in more detail below that this position is not so exalted as it seems to be. The assumption that the species category determines the systematic work with relation to a very distinct group of relationships, and thus to a determined scientific system, is false. But first we must determine whether one of the many possible and necessary systems (as discussed above), whose postulation generally is the function of the several biological disciplines, appears logically so favored that it is a priori destined to be the "general reference system." Is there, in other words, any one system that must be regarded as the task area of biological systematics?

THE PHYLOGENETIC SYSTEM AND ITS POSITION AMONG THE POSSIBLE AND NECESSARY SYSTEMS IN BIOLOGY

In the years immediately preceding World War II the relationship between the so-called idealistic morphology and phylogenetics was debated particularly passionately in the German biological literature. Naturally, the discussion was not restricted to the German literature, and it persists with basically unchanged arguments on both sides even now (cf. the reports of K. Günther, 1956, 1962).

Although not always expressed in these words, the essentials of this dispute can be reduced to the brief formula: Should the system of animals and plants developed by systematics be phylogenetic or not? Following the ideas presented in the first section of this book, the answer to this question cannot simply be that (depending on the viewpoint of the opponents) either a phylogenetic system or a system unrelated to a definite phylogenetic viewpoint is the only possible, or even the only necessary, system in biology. Rather we must emphasize that the whole dispute is justified only if it asks not for *the* biological system, but for the one among the many possible and necessary biological systems that should be selected as the general reference system. Thus the above question would more correctly be: Is the phylogenetic or a definitely nonphylogenetic system (e.g., an idealistic-morphological system) better suited to serve as a general reference system, or does one of these systems for intrinsic reasons demand this precedence over all others?

Seen from this standpoint, the extremists of both sides are wrong from the outset in denying the right to existence to any system other than the one supported by them. Such an interpretation is presented, for example, by Zündorf (1942): "Either organisms have evolved, in which case a type doctrine and idealistic morphology are superfluous, or the 'types' owe their existence to supernatural (idealogical) events, in which case there could be no phylogeny." Even von Wettstein reviews a work of Zenkewitsch with the words: "He correctly emphasizes that only a phylogenetic system is possible."

Even if, as we will try to show, a phylogenetic system is to be preferred among all conceivable biological systems, it is still necessary to erect other systems, even purely idealistic-morphological ones. By "phylogenetic system" we mean a system that expresses the phylogenetic relationships of organisms. We will investigate below, in great detail, the concept of "phylogenetic relationship," along with everything related to it. For the present it is enough to say that the "phylogenetic" relationship has the character of genealogical relationships between organisms and groups of organisms. Genealogical relationships are, however, something entirely different from "similarity." It is usually far more difficult to determine them than it is to discover correspondences and differences in the form (holomorphy), mode of life, and distribution of organisms. In the eyes of many biologists the phylogenetic system is therefore clearly at a disadvantage compared to idealistic morphology and systematics (for which according to Zündorf, 1940, the names typology, *bauplan* research, ideal-developmental morphology, pure morphology, traditional and classical morphology have been used), because it is believed that a purely morphological system (in which the individual organisms and the categories that result from the grouping of organisms on the basis of the structural plan of their bodies appear as expressions of variations of increasingly inclusive types of structural plans) precedes logically or at least historically a phylogenetic system.

"Idealistic morphology has been the prerequisite for introducing phylogenetics not only in the history of science, but even today must precede it on logical grounds" (Naef, 1919). The same author stated in 1931 that "phylogeny is basically only a kind of translation of the results of systematic biology, especially of morphology, into the language of the theory of descent, and not a special field of research." Finally, the botanist Troll (1940) presented the view that "the doctrine of descent . . . arose historically from the type concept of comparative morphology," that phylogenetics must be based on the typological groupings achieved by idealistic morphologists, and that typological methods are indispensable to the doctrine of descent because the historical roots of phylogenetic thought lie in idealistic morphology (cited from Zündorf 1940, and Zimmermann 1943).

Related to this concept is the differentiation of steps into which various authors divide the development of taxonomic work. Klingstedt (1937), whose work was reviewed with particular approval by Theil (in Fortschr. Zool.), distinguishes three steps:

(1) Description of species and more or less arbitrary arrangement—classificatory stage;
(2) Ordering of species according to all their characters with the object of a typological system;
(3) Phylogenetic stage.

Zimmermann, one of the best of modern theoreticians of systematic work, expresses himself similarly: "Each individual treatment of groupings consists . . . of two portions. In the first ('phytographic') portion the plants, parts of plants, etc. are studied comparatively as individual plants, etc. In the second (true grouping) part the grouping itself then follows on the basis of the individual phytographic studies (Zimmermann 1937, apparently following Diels 1921).

The idea of the nature of systematic work expressed in this succession of steps seems particularly well founded because it agrees closely with the usual interpretation of how biological insight in general comes about. According to this interpretation, the first step is ordering, the simple and comparative description of objects. In the second step the causal, holistic, and historical relationships are investigated and—in general biology—rules are formulated for the uniformities of phenomena that appear. The third step, finally—theoretical biology—develops, with the help of theoretical assumptions, laws of biological phenomena (von Bertalanffy 1932).

In view of this state of affairs it is not surprising that many systematists have concluded that in biology a system constructed according to the principles of classification, or at least of typological steps (in the sense of Klingstedt's classification), must occupy a favored position as being least hypothetical and closest to the "facts." Holdhaus apparently represents this interpretation when he asserts that the "true systematist" has to classify; only secondarily can he enter into phylogenetic speculation (cited from Karny 1925). Horn (1929) believes the systematist has to order, without saying anything as to how the ordering he presents came about, and (1929) the "smaller epigones" of Linnaeus first coupled "systematics with phylogeny." According to Börner (1929) the systematist orders "into arbitrary relations according to degrees of similarity. The phylogenist checks the systematist's degrees of similarity for phylogenetic homogeneity, and constructs the missing intermediate members in order to restore the phylogenetic connections between existing types. Thus do critical phylogeny and systematics best support one another. The former is the criterion for the latter, and to a certain extent represents a higher level of systematics." Even Zimmermann, one of the most zealous of modern advocates of a consistent phylogenetic systematics, still regards systematics and phylogeny as two different things—apparently in regard to his separation of work on groupings into steps, as mentioned above. Zimmermann (1937) criticizes the Königsberg phylogenetic tree of plants of Mez in the words "This presentation of living organisms 'above' or 'below,' 'near the main stem,' or 'far from the main stem,' etc. is a systematic, but not a phylogenetic, arrangement."

The idea expressed in all these utterances, that phylogenetic systematics is based logically and/or historically on purely morphological or at least nonphylogenetic systems, and the view often derived from this idea that a pure (idealistic) morphological or at least nonphylogenetic system therefore merits precedence over the phylogenetic one because it stands closer to the natural facts and contains fewer hypothetical elements, is absolutely wrong. It is based on two false assumptions. One is that it is possible to carry on science (including systematics) without making assumptions. The participation of human activity in the process of perception, so strongly emphasized (see for example Gehlen 1940 and 1942; in the same sense Wein 1942 speaks of "the new 'actional' structure of the subject-object designation"), must make us suspicious of the simple notion that the first step in systematics could be a simple "classifying" without regard to the backgrounds of the ordering expressed in such classification. In reality any ordering and "classification" consists of a consideration and presentation of natural reality from a certain point of view. If the systematist approaches his task without clear knowledge, and therefore without a consistent viewpoint, the result is in no way a nonhypothetical system that especially approximates the facts of nature. With such an attitude the varying distinctness with which the relations between natural things are conspicuously visible, changing from case to case, rather causes a continuous changing of viewpoint, resulting in a "system" in which the relations between natural things appear to be projected through one another in such a tangled way that this picture is completely useless for solving problems, for gaining an understanding of the nature of the diversity of nature.

The second error in the idea of a logical and historical primacy of nonphylogenetic (e.g., typological) systematics stems from the assumption that in biological systematics the primary relationships between living entities are similarity relationships (especially those of structural morphology), and that the elements of systematic work should be individuals, if not even species.

"In nature we find only single individuals of animals and plants. In the same way that man deals with the rest of his environment, so he has grouped animals and plants according to their similarities and designated the resulting groups with special names." In contrast to this interpretation, we have tried to show that it is not individuals, but semaphoronts, i.e., individuals in a definite, very brief interval of their life span, that must be regarded as the elements of systematics. The brevity of the interval that in many cases suffices to change an individual from one form phase to another—or, to express it more adequately for our purpose, the possibility of observing directly the genetic relations between semaphoronts that differ in form—must necessarily result in genetic viewpoints being considered in systematics from the beginning. They are present even in the oldest and traditional classifications of animals.

According to Arnold (1939), already in 1550 B.C., Eber's Egyptian papyrus recognized the development of the scarab from the egg, the fly from the maggot, and the frog from the tadpole. It would not have occurred to any of the ancient systematists to classify differently, because of morphological differences, the young and adult stages of animals, insofar as their genetic relationships were known to them. Anyone who believes that the "species" is the unit of the biological system (e.g., Plate 1914, p. 119; Collier 1924, p. 640; Börner 1929, p. 72) must first properly recognize that genetic viewpoints were considered even in the ancient classifications of animals, for (as we will show more precisely, and as the simple fact of metamorphosis already shows), genetic criteria cannot be divorced from the species concept. Möbius (1886) mentions that John Ray, one of the best known predecessors of Linnaeus, pointed to the genetic criterion as "a sure sign of specific correspondence, and he did so with clear scientific consciousness."

Consequently the development of biological systematics was not simply a replacement of an originally incomplete morphological system by more and more complete morphological systems, from which the phylogenetic system then developed by simple reinterpretation of morphological similarities in terms of phylogenetic blood relationships. Rather the older systems must be viewed as combined constructions that presented the most apparent relationships among organisms, even though these relationships belonged to entirely different dimensions of the total construction. These old systems already contained the germs of all later biological systems, and the later systems developed from the earlier in the true sense of the word.

That ecological, physiological, and psychological criteria were also used in the older systems is suggested by the fact that they opposed man to all other organisms. This is certainly not justified on morphological grounds, but only by an evaluation of man's psychic characters (in the broadest sense). That this was in fact the reason is shown by the Linnean designation *Homo sapiens*, although Linnaeus himself did not oppose man to all other animals. It would be the task of a comprehensive history of biological systematics to determine which among the relationships existing among organisms were taken into account in the older systematics. Here we may merely point to such groupings as the "blood animals," the "bloodless animals," and the "egg-laying quadrupeds" in Aristotle. The history of ornithology of Stresemann (1951) shows, at least for a small group of animals, which points of view were decisive in the construction of the older systems of birds.

We do not maintain that, in the course of their historic development, the various systems that were contained germlike in the older systems were all equally justified and have taken their place independently side by side. Rather the ecological, physiological, etc. viewpoints have been mostly excluded in the work of grouping. Thus in the science of systematics attention has been focused more and more on morphological—and finally phylogenetic—systems. The ecological, physiological, psychological, chorological, etc. relationships have been treated differently, and so far no especially elaborated systems of these relationships have been formulated. In this sense the view of Klingstedt and others of a steplike development of systematic work certainly contains much that is correct. But our view seems to me to have decisive advantages over others in evaluating the logical position of biological systematics and its present and future tasks.

Points of view have not been clearly separated even today; the most modern systems still contain juxtaposed groupings based on different viewpoints. Even the most critical and most modern theoreticians of systematic work (e.g., Plate 1914, Diels 1921, Zimmermann 1931) regard "combined systems" as "particularly valuable" (Plate) or at least productive (Zimmermann). This is attributable essentially to the fact that the principles on which scientific systematics must be based, even in biology, have not yet been debated in a really adequate way.

Historically the mixing of morphological and genetic criteria has been significant. The fact that genetic criteria were already used in the older systems does not mean, of course, that it was realized from the beginning that there are genetic relationships between all organisms and all semaphoronts. But the ontogenetic relationships of the semaphoronts, and the obvious genetic relationships between the individuals of a species, are in reality only partial phenomena of the hologenetic relationships that interconnect all semaphoronts. Therefore the inclusion of even only certain genetic criteria in biological systematics necessarily resulted in the realization that there are in fact genetic relationships between all organisms. The theory of descent, that is the perception that the existing diversity of life on the earth arose historically from an earlier simpler condition, and that the semaphoronts—the elements of all systematic efforts in biology—must be regarded among other things as members of a community of descent, is thus derivable from biological systematics (not "morphological systematics"). This is, in fact, its most important result to date.

The historical relations between the theory of descent and biological systematics are of two kinds. On the one hand, different semaphoronts can be connected with one another within the life cycle of an individual by directly observable genetic relations. On the other hand, different semaphoronts can enter into relationships with one another for the purpose of producing common, but under certain circumstances clearly different, offspring. Since these and other important facts group themselves around the concept of the biological "species," it is accurate up to a point to say that the theory of descent is a result of a more profound study of the species category (Hertwig 1914). The statement by Dingler (1929) that "the remarkable phenomenon that the genetic idea was conceived so late in human thought, especially that it does not appear in Greek natural philosophy, is related to the absence of a clearly formulated species concept" is probably correct in principle.

This study could become fruitful in the sense of the theory of descent, however, only if another historically important fact is taken into account. This is the fact that the closer and broader relationships of morphological similarity among all the species can often be best expressed if the species are gathered into a hierarchic system of groups. It is doubtless evident from the observation that an "organism" can change in a relatively short time from one form in which it resembles certain organisms into another form, dissimilar to its earlier condition, that makes it more similar to other organisms, and from the further observation that the offspring of two different individuals may be dissimilar to their parents, to conclude that even greater changes in form may possibly take place in the course of periods of time so vast that they cannot be surveyed. Only a small, though decisive, further step is needed to conclude that the graduated differences in form that are expressed in the hierarchic system of species originated by similar processes of change lying farther back in time or requiring longer periods of time. Thus one has the most essential declaration of the theory of descent.

In comprehensive accounts of the theory of descent a great number of "proofs" of its correctness are generally given (paleontological, embryological, zoogeographic, and others). All these proofs are undoubtedly significant, but it must be pointed out that they gain this significance only through their relation to the hierarchic system and the similarity relationships (which need not be morpho-logical in the narrow sense) of the semaphoronts that are expressed in this sys-tem. The fact that semaphoronts (-complexes) which, because of their similarity relationships are included in the same group, may prove to be connected by other entirely different relationships (zoogeographic, for example) that were not taken into account in the original compilation calls for an explanation. The ex-planation is then provided by assuming common descent, by inferring that com-munity of similarity = community of descent.

To many biological systematists the hierarchic system appears to be the only possible system, or at least the most important one, especially because it is often preferred where, among several possible systems, none seems to be superior for intrinsic reasons. That this is not really true is shown by the periodic system of the chemical elements, and the 32-class system of crystallography. The periodic system shows the importance of discovering the system that is best suited to the scientific investigation of the relationships among a particular multitude of indi-vidual objects. Bavink (1941) calls the erection, or better the discovery, of the periodic system "one of the most fundamental and important results of all nat-ural science."

In biological systematics repeated attempts have been made to introduce sys-tems other than the hierarchic form. Plate (1914) mentions as an example the ornithological "quinary system" of Kaup (1849), "which is based on the conten-tion that all differences and groupings of vertebrates appear according to the number five." Similar quinary systems have been proposed by other zoologists (Oken, MacLeay, Vigors). Reichenbach proposed a quaternary system. Further examples are given by Stresemann (1951). A reproduction of MacLeay's quinary system is shown in Fig. 1, taken from Wilson and Doner's review of the systems of insects that have been proposed. Plate is probably correct in attributing to a similar "feeling for symmetry" the opposition of many biologists to "making the Passeres with more than 6,000 species equivalent to the ratites with only 20 spe-cies, or subdividing the latter group even further." According to Paramanow (1937), Lubistshev has occupied himself with the possibilities of other biological systems that deviate from the hierarchic system. "Lubistshev calls one such sys-tem 'combinative.' It has the appearance of a crystallographic latticework of many dimensions corresponding to the number of independently varying char-acters. The characters are all of equal value. The latticework of many dimensions graphically reflects this system. According to Lubistchev, the third most com-plete system is the correlative system. The periodic table of the elements is to be regarded as a special case of the correlative system. It can be represented graph-ically as a spiral around a cylinder: in this case either one character or a small group of characters have a dominant significance, and the other characters stand in a correlative relationship to them." A "periodic system of butterflies" has even been attempted (Bachmetjev 1903-04)! As will be shown below, it is no accident that practically all authors who recommend a nonhierarchic system also support some kind of nonphylogenetic, mostly typological (idealistic-morphological) systematics. The fact that none of the various nonhierarchic systems has survived suggests that the hierarchic system best expresses the structure of the complex of relations that interconnects all organisms. Consequently we must now examine it more closely.

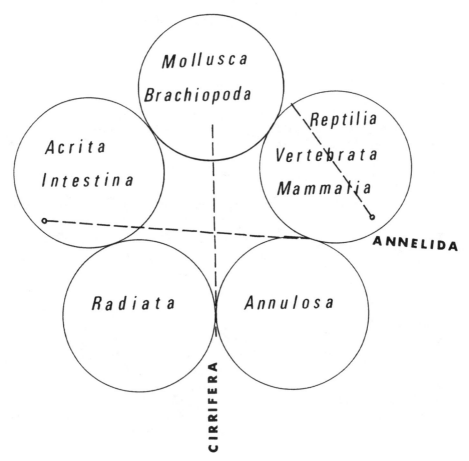

Figure 1. Quinary system of the animal kingdom according to MacLeay. (Redrawn from Wilson and Doner.)

The hierarchic type of system has been closely studied by Woodger. Gregg has presented the results of these investigations in his book *The Language of Taxonomy* (1954). This work is done with the methods of symbolic logic. We consider the investigations of Woodger and Gregg extraordinarily important because they clarify, with methods that exclude all confusion and contradiction, the peculiarities of the hierarchic system, and so create exact prerequisites for investigating the questions of whether and why it deserves the favor it enjoys in biological systematics. According to Woodger, the definition of the "hierarchic system" is: $HR =$ the set in which any relation z is a member if and only if ($z \in$ one-many) and (z' ' $UN = \check{z}_{pu}'$ ' $\{B'z\}$. The content of this definition may also be presented graphically (Fig. 2). It states that the elements ($x_0, x_1 \ldots x_9$)

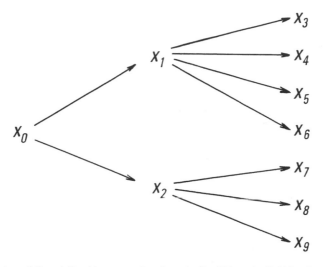

Figure 2. Structure of the relationships among the elements of a "hierarchy." (After Gregg.)

composing the ordered quantity in the system are paired by relations that extend in only one direction. Such relations exist, for example, between mother and child, father and son, employer and employee. In Fig. 2 these relations (z) are represented by arrows. In quantities that are organized nonhierarchically there are also relationships that are not unidirectional, as for example those between brothers. Here the elements are freely interchangeable. Such relationships do not exist between two elements in a hierarchy. If, therefore, the relationships between the elements of a hierarchy are represented by unidirectional arrows, then according to Woodger's definition: (1) The point of one, and only one, arrow can lie in each element of the hierarchy, whereas several arrows may arise from it. (2) There is one, and only one, element from which arrows emanate but to which no arrow leads. Woodger and Gregg call this element the "beginner." (3) All elements to which an arrow leads, and which therefore lie at an arrow tip, are connected with the beginner by an arrow or a sequence of arrows.

If these conditions are fulfilled by a quantity, then according to Woodger one speaks of a hierarchy. The investigations of Woodger and Gregg deal only with the formal side of the problem. At first the nature of the elements in Fig. 2 that are characterized by x_0, x_1 . . . x_9 remains completely undetermined. In order to answer the question of whether the hierarchic system is rightfully used in biological systematics we must investigate whether semaphoronts can be substituted for x_0, x_1 . . . in Fig. 2. Obviously they cannot. It is true that the individual semaphoronts are connected to semaphoront complexes (which we call individuals) by relations that we call ontogenetic relations. But the structure of these ontogenetic relations does not correspond to the conditions of a hierarchic system. There are also genealogical relations between individuals, and the structure of these relations is determined by the mode of bisexual reproduction. If we could investigate the origin of every living individual found in nature, we would find that in the vast majority of cases it descended from two individuals of different sex, its parents. Since several individuals often arise from the same parents, and the parents themselves arose from pairs of parents, we would find a great number of individuals interconnected by such genealogical relationships. But not all individuals are interconnected in this way. If we could determine the genealogical relationships among all individuals over a long period of time, and present these relationships graphically, we would find gaps in the structure of the relationships. These gaps divide complexes of individuals, which we call species, from one another. Despite all the difficulties that the species concept poses in detail, it is an established fact that these species do in fact exist in nature, and that they are

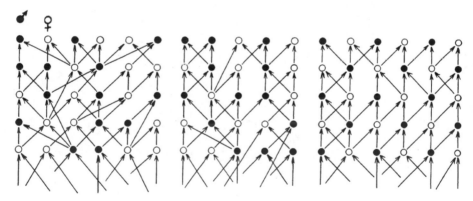

Figure 3. The genealogical relationships theoretically existing among the individuals of three species.

relatively stable, reproductively isolated, complexes. They are complexes of individuals interconnected by genealogical relationships like those shown in Fig. 3, individuals capable of producing new individuals unrestrictedly only among each other.

As Fig. 3 shows, the structure of the genealogical relationships within the species does not correspond to the definition of a hierarchic system any better than does the structure of the ontogenetic relationships that connect the individual semaphoronts with one another. Consequently we cannot substitute individuals for the symbols x_0, x_1, x_2 . . . in Fig. 2 any more than we can substitute semaphoronts.

Species are relatively stable complexes that persist over long periods of time, but they are not absolutely permanent. Modern genetics tells us how new species

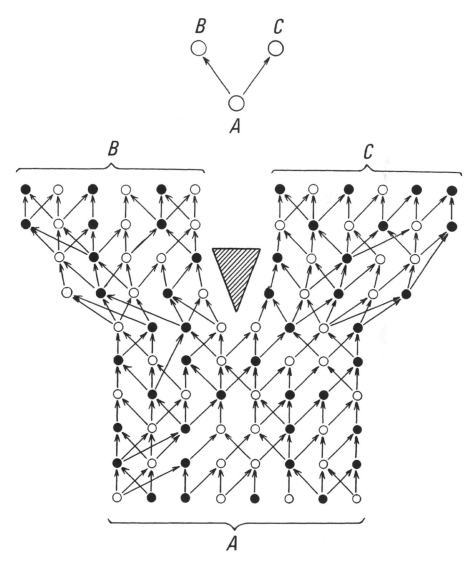

Figure 4. The process of species cleavage.

arise (Fig. 4): new gaps arise in the genealogical relations as a result of reproductive isolation. These gaps delimit new species. As Dobzhansky and many
others assume, geographical barriers play the chief role in this isolation. From
these empirical facts regarding the origin of species we may assume that the gaps
between existing species arose in the same way at some time in the past. This
means, therefore, that we would find the gaps separating existing species bridged
over if we could go back into the past. If, for example, we follow backward the
stream of genealogical relations among the individuals of species B, and do the
same for species C, at some time in the past we would arrive at one and the same
species A, which we would call the common stem species of B and C.

We will call "phylogenetic relationships" the genetic (genealogical) relations
between different sections (in the diagram, Fig. 4), each bounded by two cleavage processes in the sequence of individuals that are connected by tokogenetic
relations. They are represented by lines or arrows in Fig. 4. The structure of the
phylogenetic relationships that must exist between all species according to the
assertions of the theory of descent is necessarily that shown in Fig. 2. It is evident that this structural picture of phylogenetic relationships agrees exactly with
the structural picture of the relations that must exist between the elements of a
quantity if the quantity is to be called a hierarchy according to Woodger's definition. Therefore the species of biological systematics can be substituted for x_0,
x_1, x_2 . . . in Fig. 2. We can prove this by comparing Woodger's definition point
by point with what we know of the structure of the phylogenetic relationships
between species. But this means only that there is a firm relation between the
hierarchic type of system, which has been used almost exclusively in biological
systematics, and a certain kind of relationship among organisms: the hierarchic
system is the adequate form of representation for the phylogenetic relationships
between species.

Consequently the question arises whether the fact that biology has held unswervingly to the hierarchic system in the face of the many attempts to introduce
nonhierarchic systems is of deeper significance, and whether the phylogenetic
system, which corresponds exactly to this hierarchic type, is not indeed the general reference system of biology that we are seeking—despite the historical and
logical primacy claimed for morphological systematics.

It is probably of great historical significance that Darwin himself expressed
the thought that the possibility of arranging organisms in a hierarchic system is
explainable only by assuming a phylogenetic relationship among them: "the
simple fact that species, both extinct and living, break down into genera, families,
orders, etc.—a division analogous to that underlying varieties—would otherwise
be unexplainable, and seems unremarkable to us only because it is commonplace."
Uhlmann (1923) adds that "Darwin was well aware that expressions such as
'plan of creation,' and 'unity of type' do not explain this possibility of subordination, but that only common descent can help explain this marvel." Similarly, Hertwig (1914) says: "Finally, the extremely characteristic arrangement of the circles

of relationship that confronts us in the animal and plant kingdoms is important evidence favoring the theory of descent. . . . The relationships of the species in a genus are most suitably expressed by the graphic presentation used in human genealogy, namely the family tree. The family tree is also the only form that adequately expresses the anatomical relations of the individual animal phyla." The "phylogenetic tree" is only a different, sketchlike form of presenting the hierarchic system.

With respect to the logical significance of the hierarchic system, recent attempts have been made to base the theory of descent—waiving the various historical "proofs" that led to its general acceptance—solely on the fact that the similarity relationships between organisms or (more accurately) species can be best represented by arranging them in a hierarchic system. As Günther (1956) says, this was done most consistently by Tschulock (1922).

It is true that a hierarchic system of relations can be created arbitrarily by man. In the organization of a governmental authority or of an army, the authority of the different functionaries may be so arranged that the relations between them correspond to those of a hierarchic system. The designation itself (hierarchy means rule by priests) is derived from an institution created by human will. An identification key also has the character of a hierarchy, and here too the choice of characters that, hierarchically subordinated, lead to the different species is determined entirely by the discretion of the author.

But if there are relationships between natural bodies that obviously were not instituted by man but whose structure corresponds to that of a hierarchic system, then the only acceptable explanation for the occurrence of this structure seems to be the assumption of a "hierarchy of partition" (Woodger, see von Bertalanffy, 1931). Authors who have recognized this have accused phylogenetic systematics of circular reasoning. Günther (1956) formulates the objections of Steiner (1937), Danser (1950), Troll (1951), and others as follows: "Phylogenetics grew out of the morphologically based 'natural system,' and has to work, now as then, with morphological criteria. The derivation of ideas concerning the kinship of organisms via natural affinities determined by phylogenetics is a vicious circle, a *petitio principii*."

In reality, phylogenetic systematics uses a method known and employed in all sciences, which in the humanities is called the "method of reciprocal illumination" (checking, correcting, and rechecking of the Anglo-Saxon authors). Mühlmann (1939) says concerning the use of this method in ethnology: "The nature of culture and its individual systems can be determined by the method of reciprocal illumination of the part in relation to the whole. The objection that this is circular reasoning has been discussed by Dilthey; it applies logically, but not to practical investigation. Juxtaposition of parts to the whole presents the practical possibility of advancing, in a succession of mental acts, to successively higher points of view. A field investigator trying to understand life in a Melanesian village does not look for a mystical approach to the culture as a whole. He studies the individual elements 'one after another,' say arts (T), social organization (S), economics (W), religion (R) (we are intentionally schematizing), and from these cultural systems he obtains preliminary rough ideas T_1, S_1, W_1, R_1. By trying to understand the functional relationships of these (the influence of arts on economic conditions, that of religion on social organizations, etc.) he obtains a preliminary rough idea of the culture, K_1. Then he lets light from K_1 fall back on T_1, S_1, W_1, and R_1 and obtains the more sharply defined ideas T_2, S_2, W_2, R_2. The functional relations of these, in turn, throw a sharper light on the culture, K_2. The light falls back again from K_2, and the ethnologist obtains T_3, S_3, W_3, and R_3, and so on."

Phylogenetic systematics operates in a very similar way. Here the phylogenetic affiliations are, so to speak, the whole. With its parts—the relations of morphological, ecological, physiological, geographic, etc. similarity, which reflect the phylogenetic affiliations of their bearers—it is subjected to the method of checking, correcting, and rechecking. With basically the same argument, Wenzl rejected the contention of the idealistic morphologist Steiner (1937), who called explaining internal affinities by means of a family tree "a circle, a *petitio principii.*" "Circular reasoning is at any rate exaggerated. The theory of descent is a hypothesis, and in any hypothesis the facts are explained by an assumption, which was itself made on the basis of the facts" (Wenzl 1938).

Consequently we can agree with Günther's rejection of the apparently baseless objections to phylogenetic systematics: "The empirico-critically constructed 'natural system,' as an encaptic system of graduated diversity of form, served as proof for the theory of descent. And only the theory of descent, in turn, uses scientific, empirico-critically based ideas to explain the encaptic diversity of form expressed by the 'natural system.' But as a form that was found by means of 'generalizing induction' (see Hartmann 1948), intended to make empirical findings understandable, the theory naturally precedes this understanding as 'a hypothesis with whose aid one tries to interpret the empirical material' (May 1949). The objection that it is a *petitio principii* does not concern this state of affairs at all, even though the phylogenetic interpretation of the empirical material works with morphological —as well as physiological, ecological, ethological, and chorological—criteria" (Günther 1956).

Having shown that the objections to the phylogenetic system are invalid, we face the question of whether there are reasons that expressly favor the phylogenetic system as a general reference system for biological systematics. Such reasons are:

(1) Making the phylogenetic system the general reference system for special systematics has the inestimable advantage that the relations to all other conceivable biological systems can be most easily represented through it. This is because the historical development of organisms must necessarily be reflected in some way in all relationships between organisms. Consequently, direct relations extend from the phylogenetic system to all other possible systems, whereas there are often no such direct relations between these other systems.

(2) The kinship relations between groups of the phylogenetic system are meas-

urable, at least in principle, and in an exactly defined sense. On the other hand, the form relationships between groups of morphological systems in the broadest sense are not exactly measurable, since there is no known method of measuring similarities and differences in form. Many believe the mathematical bases for exact measurements of similarities and differences in form will eventually be created, but we are skeptical. In any case we have to work with what is presently available, and with these tools it is impossible to construct any kind of morphological system on an exactly measurable foundation.

(3) In many, perhaps most, cases the similarity relations of organisms (particularly the morphological ones) are identical with the relations of phylogenetic kinship. As a result the groups of the phylogenetic system in most cases are also the groups of closest form relationships (similarity). Even in cases where phylogenetic relationships and form similarities are not identical (as where the structure of the similarity relationships is reticular), the groups of the phylogenetic system are usually relatively easy to recognize, by suitable evaluation of the various similarities, through characteristic features or at least combinations of characteristic features. Consequently the phylogenetic system has the practical advantage that its groupings are also relatively easy to characterize morphologically. To be sure, morphological systems have the same practical advantage insofar as they consider easily recognizable form properties of the semaphoronts. But they are less suitable as general reference systems for biological systematics in that any groups based on similarities of form have no significance beyond the purpose of the particular morphological system if they are not identical with certain groups of the phylogenetic system.

(4) The attempt to introduce a system based exclusively on morphological viewpoints or on morphological form similarities (in the broad sense) as the general reference system of biological systematics would encounter serious difficulties. Stages of metamorphosis, polymorphisms, and other semaphoronts that differ in form but obviously belong together genetically, would have to be grouped without regard to these genetic relationships (e.g., caterpillars would have to be placed in one group, and butterflies in an entirely different group). To avoid this it would be necessary to give absolute precedence to a peculiarity of the individuals that is only conditionally to be recognized as a "form character"—namely, "the capacity under certain circumstances to alter their form in a certain way, for example in a particular cycle of metamorphosis"—without being able to do anything comparable in the higher group categories. In discussing morphological systems no one would think of recognizing "the capacity to alter its form in a certain way under certain circumstances" as a morphological character.

The points mentioned above will be discussed later in more detail. Purely theoretically, point (1) is decisive since it shows that the choice of a general reference system for biological systematics is not at all free, but for intrinsic reasons must be the phylogenetic system. We may assume that the authors who were criticized above (p. 9) for asserting that only a phylogenetic system is possible or justified meant it in this sense. The reasons for "choosing" the phylogenetic system that are given in (2) to (4) are merely intended to show that the phylogenetic system has other advantages besides its logically preferred position.

In view of the great importance of our subject I consider it useful to present a summary of the preceding discussion. The following seems to me a particularly graphic form of presentation.

We may start from the concept of life as a multidimensional diversity. This means that we are confronted with an immense number of individual organisms that differ in many respects (directions, dimensions). For a graphic presentation (Fig. 5A) we can think of the dimensions in which the differences lie as comprising three main directions. These we will call the morphological, physiological, and psychological dimensions. This also explains the subjects of the several biological disciplines if we imagine morphology as investigating the relations of the organisms that run in the morphological dimension, physiology those in the physiological, and psychology those in the psychological dimension. If we remember that the three main directions of our presentation really subsume several neighboring dimensions, then the position of the various subdisciplines of morphology (cytology, histology, organology, etc.), of physiology (stimulus physiology, metabolism, etc.), and psychology also becomes clear. But we must also consider that the relations interconnecting the multidimensional diversity of organisms extend not only in the planes of the several dimensions that are distinguished. There are also others that interconnect the different dimensions. There are important relations between the body form, the vital functions, and the behavior of individual organisms as well as of entire groups of organisms. Consequently there is the further possibility of choosing a point of view between the major dimensions and considering just the relations that extend between the major dimensions. This gives rise to disciplines that are often regarded as particularly modern, such as histophysiology or biological anatomy.

It is possible to adopt an almost unlimited number of points of view, all of which have their significance, and thus to create an almost unlimited number of new biological disciplines. It must be admitted, however, that they would not all be equally relevant.

The peculiarities of body form, vital functions, and behavior are not the only major directions in which organisms differ from one another. All three are bound to the form (the total form, or holomorphy) of the organism. Organisms also differ in their position in space, and in the time of their origin as individuals and groups of individuals. Consequently we can broaden our picture (Fig. 5B). But we must keep in mind that, particularly in the chorological dimensions, a great many individual directions have been subsumed. This is because the environment in which the organisms are distributed is itself multidimensional (see p. 47). In the holomorphological dimensions of Fig. 5B we have subsumed characters that in Fig. 5A were distinguished as morphological, physiological, and psychological.

Zoogeography, ecology, and genetics (in the broad sense) are disciplines that have chosen a position within these other major dimensions that have been distinguished besides the holomorphological dimension.

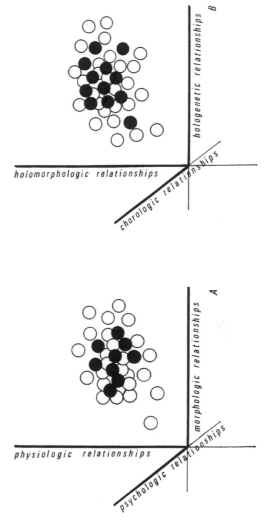

Figure 5. The position of systematics among biological disciplines dealing with the diversity of life. Explanation in text.

Now we can understand the position of systematics in the genetic dimension. From this picture we can also see that phylogenetics is not limited to one sector of the relations that interconnect organisms. As with all other disciplines, the whole structure of these relations is accessible to phylogenetics and constitutes its area of investigation; but as with any other discipline the particular forms of relationships appear in a very definite perspective. Finally, it is evident from this picture that each biological discipline must keep to its chosen viewpoint, and if it is to be cognitively incontestable and achieve results that can be used by other disciplines, it must project onto its chosen and accurately defined viewpoint the totality of the relations that fall within its scope. But it is equally evident that, in order to obtain the clearest possible survey of the whole, the individual disciplines are interdependent. Phylogenetic systematics also aims at contributing to an understanding of the phenomenon of life, to the extent that the nature of life is amenable to scientific investigation. The fact that phylogenetic systematics investigates the splitting of life into a multiplicity of different organisms, and tries to understand the structure of this multiplicity from the standpoint of its historical development, does not contradict the above, because the fact of the existence of this multiplicity and its historical development are as much a characteristic of life as are any of its other determinable properties.

By careful reflection we can also derive from our picture the most important difference between the phylogenetic and nonphylogenetic systems. Only the genetic dimension cannot be broken down further; it corresponds to the naturally given dimension of time. In this dimension both the ontogenetic relations of the different life-stages of the individual (the semaphoronts), and the genealogical (tokogenetic) relations of the individuals and the phylogenetic relations of the species, can be represented with complete accuracy and clarity. For the phylogenetic relationships this is done by means of a "family tree" or a hierarchic system.

Things are different with the "holomorphological" and chorological dimensions. Identification of the numerous incommensurable (i.e., not directly comparable) morphological, physiological, and ethological differences among semaphoronts or individuals and species with certain "dimensions" of a spatial system of coordinates is (except for hologenetic relations) only a picture intended to serve as illustration. Even aside from the fact that these relations of organisms are not measurable, or measurable only in certain extreme cases, indicating the exact "position" of a semaphoront or group of semaphoronts in the multidimensional character- or living-realm would require the use of a great many "coordinate values." In morphological or chorological systems the "relationship of form" (and "relationship of distribution") cannot be represented by the hierarchic type system, but only by the "multidimensional network" of lines of relationship. Measurability is a prerequisite in all cases.

The problems and methods of a special phylogenetics will be investigated in more detail below. Questions that have been raised about the conceptual or practical feasability, and even about the practical usefulness, of phylogenetic systematics will be discussed in detail at the most appropriate places. This will be the case particularly in the investigation of systematic problems in connection with the higher taxonomic categories, because many biologists who believe they have to express some kind of objection to phylogenetic systematics often acknowledge the inclusion of genetic viewpoints in connection with the lower categories without being aware of the consequences as set forth under point (4) above.

At this point one further question of a purely formal nature may be mentioned. This is the question of whether, in view of the fact established above that the task of special systematics (taxonomy) coincides with the task of special phylogenetics, it is appropriate to retain the old ambiguous term "systematics." Much can be said in favor of changing the name of a concept when the interpretation of the concept changes. Because of the "power of language over thought" the name long continues to be associated with obsolete ideas concerning the content of the concept that bears the name. The content of the concept of biological systematics today has little in common with the concept of systematics in the "age of systematics" between Linnaeus and Darwin. Various new designations have been proposed for partial contents of the concept of biological systematics: taxonomy, zoonomy, zoography, zoogony, etc. This creation of new names also has disadvantages, and so none of these names (except taxonomy, which was not newly created but taken over from other sciences) has been adopted. Consequently I deem it best not to make a further attempt to replace the familiar names.

II

Tasks and methods of taxonomy

"Taxonomists are more like artists than like art critics; they practise their trade and don't discuss it" (Anderson 1952). This remark certainly characterizes the disinclination of many systematists to engage in extensive discussions of the theoretical bases and methods of their science. But systematics is science and not art, and in every science there comes a time when significant progress is no longer possible without such discussion.

Before turning to these discussions we will make a few comments on critical objections many authors have made to the methods of biological systematics. The concept of "absolute certainty" plays a role in these objections, and because of imperfections in the methods this often leads to a rejection of phylogenetic systematics itself. It often leads to the rejection of any improvement and refinement of the usual methods as unessential, because naturally they too fail to lead to absolute certainty in the determination of phylogenetic kinship relations. In my opinion this position rests on a misunderstanding of the nature of all scientific endeavors. If absolute certainty of perceptions were a condition and sole justification for scientific effort there could be no science at all. A characteristic of any science is the "endless task" and the knowledge that its final goal will probably never be reached. This is also true of phylogenetic systematics. With the appearance of the theory of descent and the perception that there are phylogenetic relationships between organisms, their investigation and presentation as completely as possible became an unavoidable scientific task, whether one elevated the phylogenetic system to the position of a general reference system for biology or not. Many systematists seem to believe that the requirement of a phylogenetic systematics—and at the same time its justification—is that any author who for any reasons (e.g., a handbook of birds) has to present a system of a group of animals must present a perfect phylogenetic system. Since this is impossible, it is concluded that the requirements of the phylogenetic system are unfillable and therefore the system is to be rejected on principle.

This is a misconception. The phylogenetic system is a task whose goal is as unattainable as that of any other science. What we tentatively call the "phylogenetic system" of a group of organisms can consequently never be anything final. But we are justified in calling such a provisional system a "phylogenetic system" (in distinction to other possible systems) only if, with the aid of the presently known facts and methods, it can be made probable that it represents phylogenetic relationships more accurately than any other system. In other words, we are justified if in all probability our system approximates more closely than any other the ideal system that reflects the phylogenetic relationships absolutely correctly.

Critical discussion of the methods available to phylogenetic systematics, and any attempt to improve them, must be oriented toward what we know of the general structure of phylogenetic relationships. An answer to the question of whether we have adopted the right path depends on how accurately we know the goal to which this path should lead. We have defined the phylogenetic relationships we are trying to present as those segments of the stream of genealogical relationships that lie between two processes of speciation (p. 20 and Fig. 4). Thus by definition phylogenetic relationships exist only between species; they arise through the process of species cleavage. The key position of the species category in the phylogenetic system corresponds to the following: the species are, in the sense of the class theory, the elements of the phylogenetic system. The higher categories of this system are groupings of species according to the degree of their phylogenetic relationship.

From this there necessarily results a dichotomy of our task of investigating the methods of phylogenetic systematics. First we must investigate the methods of systematics in the area of the species category, and then we must ask what means we have at our disposal for erecting taxa of higher rank according to the principles of phylogenetic systematics.

TAXONOMIC TASKS IN THE AREA OF THE LOWER CATEGORIES

We have recognized the semaphoronts as the elements of biological systematics. Some of these semaphoronts are linked together into semaphoront groups (the individuals of everyday language) by genetic relationships, which we have called ontogenetic relationships. Consequently the "individuals" are not, strictly speaking, the lowest category of systematics. There are also genetic relationships between individuals, which we will call "tokogenetic relationships," that arise through the phenomenon called "reproduction." It is decisive for what follows that the bisexual mode of reproduction is the rule in the animal kingdom, for this means that individuals are usually the offspring of not one, but two individuals of opposite sex. Consequently, the structural picture of the tokogenetic relationships of individuals is essentially different from that of the ontogenetic relationships of semaphoronts. Groups of individuals that are interconnected by tokogenetic relationships are called species. The fact that the species concept as used in systematics is much more complicated need not concern us at the moment.

The species are not absolutely permanent either, any more than the semaphoronts and individuals are. New species arise when gaps develop in the fabric of the tokogenetic relationships. The genetic relationships that interconnect species we call phylogenetic relationships. The structural picture of the phylogenetic relationships differs as much from that of the individual tokogenetic relationships as the latter does from the structural picture of the ontogenetic relationships. In spite of these differences in their structural pictures, the phylogenetic, tokogenetic, and ontogenetic relationships are only portions of a continuous fabric of relationships that interconnect all semaphoronts and groups of semaphoronts. With Zimmermann we will call the totality of these the "hologenetic relationships."

This viewpoint will seem artificial to many systematists. Why not stay with the customary mode of expression and say: "individuals change during the course of their lives; they reproduce, and species are a consequence of the mode of their reproduction; finally, even these species change, and by their cleavage new species arise"?

Definite formulation of known facts is often, in all sciences, a prerequisite for the emergence of new questions, and thus for gaining new insights. In the present case the formulation we have chosen is justified by the fact that the changeability of organisms is not a primary issue in systematics. The primary issue is rather the existence of organisms as bearers of characters. The existence of genetic relationships—including the ontogenetic relationships between different character bearers—must be determined secondarily by systematics, often with great difficulty. This difficulty is evident from the many customary designations for different stages of the individual life cycle, which originally were given generic names (*Cysticercus, Leptocephalus*, and many others).

The ontogenetic and tokogenetic relationships differ from the phylogenetic relationships in a way that is of practical significance in systematic work—because of the short temporal duration, they can be directly observed. This assures a particularly high degree of reliability to the results of work on the lower categories, a degree of reliability that cannot be achieved for the higher categories. Another circumstance is even more important. In practice, the determination of genetic relationships—even where they are surely present and demonstrable by observation, at least in principle—is often so difficult or at least time-consuming and laborious that taxonomy often grasps for substitute methods, even in the lower categories where it is not absolutely necessary. Before such methods gain general acceptance their reliability must naturally be tested. This can best be done by comparing their results with those to which direct determination of ontogenetic relations leads. Thus for the lower categories it is possible to test directly the reliability of indirect substitute methods of phylogenetic (perhaps better called "hologenetic") systematics. This testing is the more important because the same substitute methods may be used for the higher categories, where taxonomy is limited to them because it is never possible to determine genetic relations by direct observation. Thus the results of testing the indirect methods supply important clues to their applicability in the taxonomy of the higher categories.

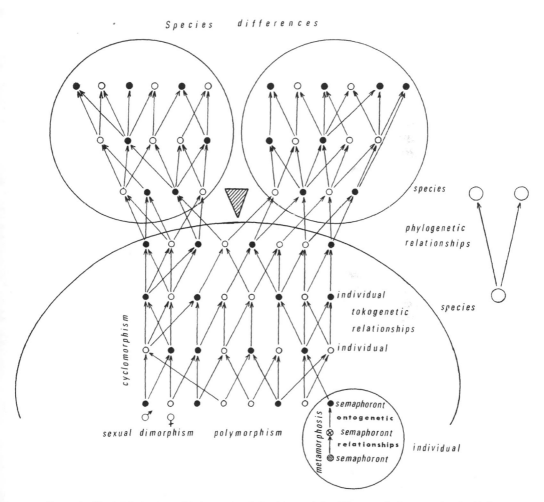

Figure 6. The total structure of hologenetic relationships and the differences in form associated with its individual parts.

The total fabric of the hologenetic relationships is shown graphically in Fig. 6. The designations commonly used for the form differences between the bearers of different categories of genetic relationships are also given. This diagram provides a guide by means of which we can investigate the significance of one of the most important tools of systematics, comparative holomorphology.

Comparative Holomorphology as an Auxiliary Science of Taxonomy: The Allomorphism of Species

GENERAL

We started from the position that the existence of species in nature is determined by the fabric of the tokogenetic relationships that exist between individuals. But in order for such a fabric to arise, individuals must have the possibility and the capability to engage in reproductive relationships. This fact is taken into account in a definition of the species that is accepted even by the majority of those authors who in general do not accept the principles of a consistent phylogenetic systematics: "a . . . reproductive community . . . is called a species. There is no other possible basis for this concept, above all no morphological basis" (Naef 1919). "Species are groups of actually (or potentially) interbreeding populations which are reproductively isolated from other such groups" (Mayr et al. 1953). Perhaps we too often forget that these are not new insights. Zimmermann (1953) quotes Buffon 1749: "For one can draw a line of demarcation between two species, that is between two groups of individuals that reproduce without blending, whereas one can unite in a single species two groups of individuals that blend during reproduction."

The possibility of producing fertile offspring is limited not only by the necessity for individuals to live together in space and time, but above all by the possession of compatible genetic constitutions. Modern genetics teaches that in nature there are no two individuals with identical sets of genes, but in each individual the genes must "cooperate harmoniously in the development of a character," since "a character is a product of the whole gene complex" (Mayr 1957).

But the genes of different individuals must also "cooperate harmoniously" when they combine to produce common descendants. This means that only individuals possessing sets of compatible genes can belong to a reproductive community. Consequently the species can also be characterized as "a reproductive community of individuals that share a common gene pool" (Dobzhansky 1950). But this introduces into the definition of a species a criterion that rests on the possession of certain bodily peculiarities. This becomes especially clear when it is recalled that the genes of the individuals manifest themselves in their holomorphological characters. "It will also be characteristic of species in general to exhibit morphological, physiological, and ecological peculiarities which are the phenotypic expression of the unique hereditary configurations present in the gene pool" (Meglitsch 1954).

This substantiates the use of comparative holomorphology, which takes into account all body-bound characters of the individuals (more accurately, of the semaphoronts), as the auxiliary science of systematics. Its discussion in the following must go back to the contention, justified above, that the simplest elements of systematics are the semaphoronts. The question that must be answered is: Are there definite criteria by which the morphological (more accurately, holomorphological) differences between semaphoronts or groups of semaphoronts can be attached to definite categories of genetic relationships (Fig. 6)?

METAMORPHISM

Differences in form between ontogenetically related semaphoronts we call metamorphisms. In everyday language, metamorphisms are the differently shaped age stages of an individual.

"We comprehend ontogenesis by fixing a series of momentary pictures or 'stages' out of an actually infinite number. In practice we select as many as seem necessary for understanding the process" (Naef). From the standpoint of systematics we actually must say that, within the limits of an individual (individual cycle), as many stages are distinguished as are practically distinguishable or as would originally have been recognized as different semaphoronts without any knowledge of the genetic relationships. The general tendency is to distinguish only a few, and to speak of metamorphosis and metamorphisms (or "stages" or "garbs" in Döderlein's sense) only if the differences are relatively great and if the duration of relative constancy of a character is appreciably longer than the period of transformation. There is no general rule for determining what constitutes a stage. As proof that differences between stages of metamorphosis are sometimes enormous it need only be mentioned that dipteran larvae have relatively recently been described as worms and molluscs. This also shows that the distinction we made between semaphoronts and individuals was well founded, and that in some cases considerable scientific systematic work is required to determine that different semaphoronts belong to the same individual cycle, or to related individual cycles of the same species.

The larva of one of our commonest fleshflies (*Sarcophaga haemorrhoidalis*) was described as a roundworm of the genus *Ascaris* by Jördens in 1802. At about the same time the larvae of the syrphid genus *Microdon* were described as molluscs (Spix 1825, von Heyden 1823, 1825, and even Simroth 1907). According to Remane (in Kükenthal-Krumbach) the worms of the small group *Kinorhyncha* were thought to be dipteran larvae by Leuckart.

The question of whether there are characters proving that two semaphoronts are stages in the metamorphosis of one individual is usually posed in practice as the question of whether there are certain kinds of characters that prove that their bearers are metamorphic stages of other semaphoronts, even when this is not otherwise evident. Are there, for example, characters that positively identify their bearers as larvae, or as imagos of metamorphic cycles? The answer is no; there is no general criterion for recognizing one semaphoront as a metamorphic stage of another. Not even maturity of the sex organs can be used. Many sema-phoronts cannot be regarded as metamorphic stages of other semaphoronts in spite of undeveloped sex organs (e.g., workers among ants and termites), and if fully developed and mature sex organs and functions are demonstrated in a semaphoront it is quite impossible to say whether different metamorphic stages preceded this condition or not. The phenomenon of dissogony, the twice-repeated occurrence of sexual maturity in the course of individual development among the Ctenophora, may also be mentioned in this connection.

Now, "the characteristic of developing ontogenetically in metamorphic stages" can be regarded as a form character of large groups of organisms, which in a certain sense can be regarded as "individuals of higher order" (see p. 122). Metamorphosis is a characteristic of the insects, the crustaceans, the amphibians, and many other groups of the most varied category levels. Within these groups, definite characters or groups of characters tend to be so characteristic of indi-vidual stages of metamorphosis that semaphoronts exhibiting such characters can be identified with certainty as metamorphic stages (larvae, for example) of other semaphoronts of a certain group of organisms (insect larvae, crustacean larvae, amphibian larvae). But even here there is no absolute certainty. Probably in every group in which metamorphosis normally occurs there are subgroups in which it is more or less reduced. Larval stages may be completely or almost com-pletely suppressed (ametaboly in insects, which is approached in the Termitox-eniidae, *Glossina*, and pupiparids), or adult stages may be completely or almost completely suppressed by shifting the reproductive function back into the larval stage (neotenic larvae, paedogenesis; *Ambystoma, Miastor, Oligarces,* etc.). These cases are of particular interest in the systematics of higher taxa, and will be mentioned again farther on. If a semaphoront, solely on the basis of our knowl-edge of morphological characters in related groups, is interpreted as a meta-morphic stage of some other semaphoront, we can never be absolutely sure that our interpretation really is correct.

The characters of metamorphic stages may fall so far outside the normal limits of the group to which they belong that, without knowledge of the factual onto-genetic relations, they could not be recognized as metamorphic stages of indi-viduals from this group. This is particularly true of metamorphic stages with parasitic or otherwise highly specialized habits. Several muscid larvae from African termite nests are placed in a special subfamily, the Prosthetosominae (see Séguy 1937). The imagos of these larvae are unknown. It can be assumed with certainty, however, that they have long been known, and that only the re-lationship between imago and larva has not been recognized because the imag-inal characters are so different from the larval characters of the group to which the imagos belong that the genetic relationship cannot be inferred from morpho-logical characters alone. We may also point out that a parasitic snail larva has been described (*Vertigo genesii* Gredl.), of which it can be said with certainty only that it belongs among the cyclorrhapids, a group of at least 15,000 species distributed among several families (Bhatia and Keilin 1937). All this shows how useless the holomorphological method may be in certain circumstances.

Of particular interest for determining the usefulness of comparative holomorphology in phylogenetic systematics is the question of whether there are identical or at least somehow comparable differences within groups of organisms in which metamorphosis occurs. In other words, does a particular qualitative or quantitative difference that has been recognized as a specific or generic difference in one stage of metamorphosis (e.g., in the imago) reappear in some form in other stages of metamorphosis (e.g., in the larva)? This question will be investigated later in discussing systematic tasks in connection with higher taxa.

For this chapter we may conclude: (1) There are no definite characters that identify their bearers as metamorphic stages of other semaphoronts. In other words, there are no larval, imaginal, etc. characters that are applicable to the whole animate world or at least to the animal kingdom. (2) The quantitative size or number of a character can in no case be regarded as indicating that a semaphoront is a metamorphic stage of some other semaphoront. (3) In circumscribed groups of organisms of various ranks there are particular characters that can be considered larval, etc. characters (thus generally as stage characters) within this particular group. But even here these characters do not have this value on the basis of their quantitative characteristics (size or number). Their validity has been established by direct observation of the genetic ties of their bearers, with the individual characters of all categories of the phylogenetic system supplying the justification for broadening the factual observations to a certain extent. There is no absolute certainty, however, even within these limits.

POLYMORPHISM

Metamorphosis was tied to the ontogenetic relations of the elements of systematics. These elements, the semaphoronts, are determined to be different units of the individual only by metamorphosis. Consequently, metamorphosis can be regarded as a systematic problem of the "lowest taxonomic units," which we have considered to be the individuals. This is likewise true of polymorphism, according to our definition of the fabric of tokogenetic relations between individuals (Fig. 6). But since this category of genetic relations characterizes the species as the group category of the phylogenetic system, polymorphism is an expression of the holomorphic differences among individuals within a species.

First among these polymorphic differences are the sex characters. Secondary and tertiary differences are distinguished beyond the primary sex characters. The degree of sexual dimorphism varies greatly and follows no general rule. In many cases the sexes are so similar that no differences can be found even between the primary sex organs, and the sex can only be inferred from the different behavior of the sex products. There are cases in which even the sex products may behave in a partly male and partly female manner (relative sexuality, see Hartmann

1943). In other cases the differences between the sexes are so pronounced and involve so many characters of the gestalt (holomorph) that it would never have been suspected, without knowledge of the factual relations, that the two sexes belong to the same species. (See examples in the discussion of the paleontological method, p. 141.)

It is possible to set up rules of limited applicability, that is for particular systematic groups of various levels, but such rules are not universally valid even in their area of general applicability. As in the case of metamorphosis, it is mostly experience that has shown particular characters to be sexually different in the groups in question. The degree of sexual dimorphism is often characteristic of larger groups of organisms in the sense that there are more and less strongly sexually dimorphic groups, but these quantitative differences always relate only to the particular sexually dimorphic characters and not to the totality of all morphological or even holomorphological characters.

Though it is not possible to specify certain characters that occur everywhere as sex characters, experience shows that differences of this or that kind between individuals (more correctly, semaphoronts) tend to be based on sexual dimorphism. In the genus *Musca*, for example, it can be assumed that if two individuals are distinguished by breadth of forehead this is a sex difference; if the primary sex organs can be shown to differ, then the difference in forehead breadth cannot be regarded as an indicator of specific difference between these two individuals if no other characters corroborate this. There are many similar rules in all groups of animals. They can be used by phylogenetic systematics for grouping purposes if the genetic relations themselves are not directly determinable. Pragmatic use of these rules can always lead to false inferences, since there are always individuals and groups of individuals that do not conform. But it is completely impossible to determine from the absolute magnitude of the differences between two individuals (thus without regard to the character on which the quantitative size difference is based) whether these individuals are sexual dimorphs of a single species or belong to different species.

As in the case of metamorphosis, we must conclude that there are no general criteria for deciding whether the gestalt differences between two individuals of different sex represent sexual dimorphism of a single species or not. This question can be finally decided only by direct determination of the presence or absence of reproductive relations. There are such criteria for limited areas of applicability, however, and their value as indicators is more or less limited even within these areas. Also they do not relate to the quantitative size of the differences (though even this may be of some minor importance), but above all to their quality; that is, in general it is not a particular quantity of differences but always a certain kind of difference that is characteristic of the sexual dimorphism in a group of organisms (of varying rank). The nature of these characters can be determined only empirically from case to case.

Within the species there are other kinds of polymorphism besides sexual dimorphism. It is long-known fact, which does not have to be supported by examples, that no two individuals of a species are completely alike. This individual variability, which is only a particular form of polymorphism, varies a great deal from species to species. There is no general rule for the extent of individual variability. For particular groups of organisms of different rank the degree of individual variability is often relatively constant within a particular order of magnitude. If one knows such groups well, it is often possible to say whether or not the gestalt differences between two individuals fall within the normal limits of variability in that group, and thus whether the individuals are likely to be the same species or not.

Among insects individual variability is particularly great in the Phasmoidea, and in the solpugids (*Galeodes*, etc.) among arachnomorphs. More important is the fact that the degree of individual variability in a species may not be equally great for all characters of the gestalt (holomorph). It may, for example, be greater for the strictly morphological characters than for particular physiological or psychological characters in the same species, and vice versa. Even morphological characters are known to vary to different degrees in different species or in species of different kinship groups. In one group the wing pattern may vary greatly while other characters vary relatively little, whereas in the most closely related group wing pattern is very constant but the bristling extremely variable, etc. If one finds individuals differing in characters that experience has shown exhibit great individual variation, perhaps even in a certain direction, in the taxon to which they belong, whereas the same individuals agree closely in characters that tend to be species-constant in the same taxon, then even without direct knowledge of the tokogenetic relations it can be assumed that these individuals belong to the same species.

Relationships between the individual variability of a species and its mode of life—or, as we prefer to say, its position in the ecological system—can be recognized. Thus among the Diptera the parasitic forms are often characterized by particularly great variability in body size. A task of "general systematics" is to follow these relationships further.

If individual variability exceeds a certain degree the aberrant individuals are often given species names as "varieties" or "aberrations." The decision as to what degree of deviation warrants giving the aberrant individuals or groups of similarly aberrant individuals the rank of a named variety or aberration is completely arbitrary. It is handled in very different ways in the several groups of the phylogenetic system. Differences in the external morphology of the individual are almost exclusively taken into account. Since many other characters besides morphological ones belong to the gestalt of organisms, it is clear that the distinguishing, and especially the naming, of varieties—as is done so ardently by amateur lepidopterists in particular—is of no essential importance to phylogenetic systematics, at least if the variants are not of importance for special purposes (e.g., genetic investigations of the inheritance of such aberrations).

This does not mean that the study of individual variability (to which aberrations and variations belong) is without significance for taxonomy. The opposite is true, since we try to find rules for the kind of correlation that exists between individual variability and the tokogenetic relationships of individuals in the various higher taxa. With the help of these rules the species limits can be determined with some certainty even when direct determination is not possible for some reason. Where the degree of individual variability is unknown and observation of the tokogenetic relationships for determining directly the species limits is not possible, investigation of the differences in form among the offspring of individual females may under certain circumstances provide information on the extent of individual variability.

In general, the differences in form among the individuals of a species group themselves around a mean, forming a unimodal curve of variation when plotted. This is the essential basis for the view that continuous variation is a sufficient

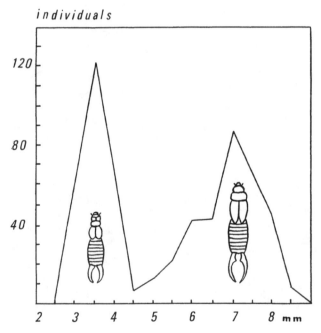

Figure 7. Bimodal curve of forceps length in male earwigs from a single group of individuals collected on islands off the coast of Northumberland. (Redrawn from Bateson and Brindley.)

criterion of individual variability, and therefore that cases of form-different individuals connected by intermediates always represent individual variants within one species, whereas individuals not so connected must necessarily belong to different species. There are also cases in which the expressions of a variable character group themselves into two or more frequency maxima, producing bimodal or multimodal curves of variation. An example of this kind is shown in Fig. 7. In this case we speak of a polymorphism of the species in the narrow sense, in contrast to the usual individual variability, but there is only a gradual difference between the two. The occurrence of such true polymorphism (which may become even more pronounced if the intermediate forms become rare, particularly if discontinuous variates, such as number of bristles, rather than continuous variates, such as length, are involved) refutes the contention (e.g., Faegri 1936) that if variability falls into two curves the splitting of this species into two species is to be assumed.

Polymorphism can often be connected with other allomorphisms. For example, it often occurs in only one sex, usually the male. In such cases it generally involves the secondary (or tertiary) characters: in species with pronounced sexual

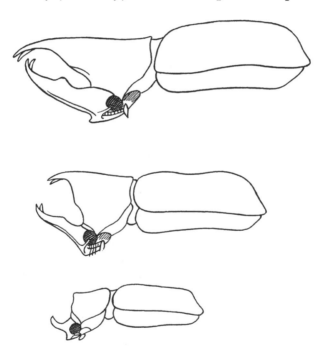

Figure 8.   Allometry in males of *Xylotrupes gideon*. (Redrawn from Bateson and Brindley.)

dimorphism, gynomorphous males often occur in addition to typical males. Here, for example, belongs the case of the European earwig that was cited above. The fact that gynomorphous males are also much smaller than normal males, of which the earwig and the beetle *Xylotrupes gideon* (Fig. 8) provide examples, indicates very clearly that gynomorphy is connected with the laws of allometric growth (see von Bertalanffy 1942). Polymorphism may be restricted to the female sex, as in the syrphid *Lasiophthalmus pyrastri*, in which three different female color phases occur, among them the var. *unicolor* (Kessel 1926). Polymorphism linked with sexual dimorphism in this way presents nothing theoretically remarkable. It is otherwise if polymorphism occurs in only one stage of metamorphosis in species in which individuals pass through a pronounced metamorphosis. In the discussion of metamorphism above it was mentioned that the question must be discussed whether, in those groups in which metamorphosis occurs, identical or at least somehow comparable subgroup differences have the same value in all stages of metamorphosis, or whether at least the degree of individual variability stands in a fixed relation in the various stages of metamorphosis. According to experience this is not the case. There are species that vary greatly in the larval stage (or in the pupal stage in insects) but only slightly in the imaginal stage, and other species in which the opposite is the case. In view of the fact that, as mentioned above, the extent of individual variability may also differ for different characters in one and the same metamorphic stage of a species, this result should have been expected from the beginning, since the characters of another stage of metamorphosis can be interpreted as a particular complex of characters of the total gestalt of the individual.

What is true for individual variability in general is also true for polymorphism in particular. Species that are polymorphic in the imaginal stage, or even in only one sex of the imaginal stage, need not be polymorphic in the larval stage. Moreover it is conceivable that there are species that are polymorphic in the larval stage but show only the usual form of individual variability in the imaginal stage. It is true that few such cases are known. An example probably belonging in this category was described by Krüger (1938), who found that in the pupa of the chironomid *Cladotanytarsus mancus* the ratio of prothoracic horn length to length of prothoracic horn bristles and the differentiation of the anal horn clearly group themselves around two frequency maxima, giving a bimodal variation curve (Fig. 9). No such bimodality of the variation curve is demonstrable for any known character in the larval or imaginal stage. On the basis of this character Krüger distinguishes a "pupal variety," *lepidocalcar*.

Thus in this example we can see a case of polymorphism in the pupal stage to which no comparable polymorphism in the other stages of metamorphosis corresponds. To regard this as "the beginning of a splitting of the species, commencing in the pupal stage," as Humphries (1937) did for a very similar case in *Trichocladius cinctus*, is apparently incorrect and at least unfounded. In reality, speciation has nothing directly to do with individual variability and polymorphism. As will be discussed in more detail below, speciation is a change in the reproductive relations between individuals. It can be, but does not have to be, linked with the origin of morphological differences. We need no more think of speciation in connection with polymorphism in the early stages of metamorphosis than we do in connection with the development of castes or with the form of

number of animals in %

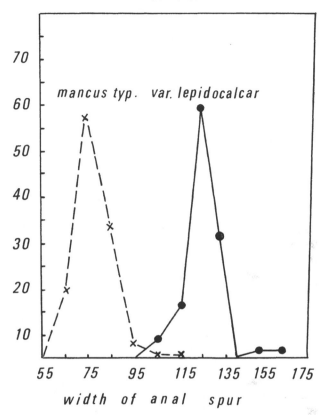

Figure 9. Frequency distribution of anal spine breadth in pupae of **Cladotanytarsus mancus.** (From measurements given by Krüger 1938.)

polymorphism, as presented above in the cases of the European earwig and of sexual dimorphism in general. In those cases in which polymorphism appears to occur in the larval stage the question will have to be investigated whether an incipient cleavage of the species is not concealed behind this phenomenon, i.e., whether there are in fact unrestricted reproductive relationships between the imagos that arise from dimorphic larvae, as has been assumed on the basis of the relatively great uniformity of the imagos. This view is presented by Emden (1958), who collected a large number of cases in which there is pronounced dimorphism in the larval stage of insects that lack dimorphism in the imaginal stage.

The fact that the degree of individual variability (including polymorphism) can be very different in the several metamorphic stages of a species, so that the range of variability in one stage cannot be determined directly from what is seen in another, is of very great importance for the applicability of the morphological method in phylogenetic systematics. Assume, for example, that two larvae, a and b, show a certain amount of difference. One of these, a, belongs to species A, which is well known in the imaginal stage and whose individual variability in the imaginal stage is well known. We cannot conclude from the fact that the magnitude of the difference between larvae a and b does not exceed the range of variability observed in the imaginal stage of species A, that larva b belongs to the same species A as does larva a. Before coming to such a conclusion the individual variability of species A in the larval stage would also have to be known. Knowledge of its variability in the imaginal stage is not enough.

We will see later on that this fact has great significance for the taxonomy of the group categories of higher rank. Here it need only be pointed out again that there is no absolutely fixed correlation between similarity of form and the nature of genetic relations. Consequently there are no general laws concerning their relationships to one another, although such correlations can be ascertained for certain areas of applicability. The extent of these areas of applicability differs greatly among the various groups of organisms. The nature of the relations and the degree of correlative connection must be determined empirically in each case.

CYCLOMORPHISM

Under the concept of cyclomorphism I include a group of phenomena that in a certain sense represent a combination of metamorphism and polymorphism. The combination of metamorphosis and reproduction results in a periodical alteration of the gestalt relations of a species. This periodicity is particularly evident in species in which reproduction is strictly seasonal. Here at certain times of the year all the individuals of a species are present only in a particular stage of meta-morphosis (e.g., as larvae). After a certain length of time they all change into the next stage, and reappear in the morphological picture of the species only after a long period of time. In many animals similar gestalt cycles are brought about by the phenomenon of rut, a period bound to the individual cycle on which an-other periodicity with longer phase length may then be superimposed. Finally, successive *generations* may become different in gestalt, and these longer-lasting alterations of gestalt, extending beyond the limits of the individual cycle, we call cyclomorphism. Depending on whether or not cyclomorphism is connected with alteration in the mode of reproduction, we distinguish between alternation of generations and seasonal dimorphism. The term alternation of generations refers only to change in the mode of reproduction (alternate asexual and sexual repro-duction in metagenesis, and alternate unisexual and bisexual reproduction in heterogony), not to the change in gestalt that often is combined with it. We may call the changes in form dimorphism or polymorphism.

Döderlein (1902) places alternation of generations and seasonal dimorphism among polymorphisms in the narrow sense, as "alternating polymorphism." This seems to me to express less well the peculiarities of the phenomenon; in particular, the relations to metamorphosis are more neglected than is desirable. For many purposes I would even consider it more correct to subsume the phenomena of metamorphosis, alternation of generations, and seasonal dimorphism under a common designation that emphasizes the periodic character of all these phenomena.

Cyclomorphosis offers nothing basically new for the problems that interest us (which agrees well with its intermediate position), so we need not discuss it further. To emphasize its intermediate position it may merely be pointed out that it can be connected with other allomorphisms of the most varied kinds, and may even shade into them. The metamorphosis of *Miastor* and other heteropezids, for example, can equally well be called heterogony, since the larvae reproduce parthenogenetically. The seasonality of the alternation of generations in aphids is a feature that is otherwise characteristic of seasonal dimorphism. In many parts of Africa, because of the slight difference between seasons and the less strictly regulated sequence of generations, the seasonally dimorphic generations of many butterflies live side by side at certain times of the year, simulating the true polymorphism that occurs in other parts of the world.

SUMMARY

The many possible transitions between metamorphism, polymorphism, and cyclomorphism suggest that they are not basically different forms of the general allomorphism of the species. This is why investigation of the correlations between various kinds of autogenetic relations of the various forms of allomorphisms leads to the same result for all. Consequently the results of our testing of the usefulness of comparative holomorphology in the task of grouping among the lower taxa are:

(1) There is neither a particular degree nor a particular kind of correspondence that justifies regarding two semaphoronts in the phylogenetic system as members of one and the same group category of lower rank. Neither degree nor kind of similarity relations between two semaphoronts permits conclusions as to the genetic relations between these semaphoronts.

(2) For limited areas of applicability—that is, in certain groups of higher rank —there are rules for correlations between particular relationships of holomorphic similarity and particular types of autogenetic relationships. But these rules concern less the absolute magnitude of the differences as the character of choice. In the Diptera, for example, there are certain types of larval characters, but not a particular characteristic degree of difference for the larva-imago relationship. This is true not only for metamorphosis and the ontogenetic relations, but also for polymorphism and the tokogenetic relations, and finally for the cyclomorphosis that exists between them. Both the area of applicability of these rules and any statement about the choice of characters and the intimacy of the correlations must be determined empirically in each case.

Within the specified limits, comparative holomorphology can (with constant attention to the fact that it can only yield probabilities) be used to solve special systematic problems. Thus rejection of a morphological definition of the species does not mean that in practice systematics has to do without morphological aids in determining the limits of species.

Of particular significance is comparative holomorphology as a taxonomic aid in those groups of organisms that do not reproduce bisexually. The genetic concept of the species, which defines a species as a reproductive community or a community of individuals capable of producing common offspring, is based entirely on the relationships in organisms with separate sexes. It can also be used without trouble for hermaphrodites with cross fertilization, but fails completely for hermaphrodites with obligatory self-fertilization (e.g., cestodes) and for organisms with exclusively unisexual or asexual reproduction. "The 'genetic species' of the new systematics has nothing of particular value for the student of parasitic flatworms." "Since the worms are hermaphroditic and self-fertilizing, and asexual generations alternate with parasitic ones, the idea of recognizing and defining genetic species is fatuous and illusory" (Stunkard 1957; similarly for parasitic Protozoa, Newell 1957). We cannot agree with Stunkard's conclusion that "it is clear that a definition which is applicable to only a segment of the animal kingdom, to existing, dioecious, interbreeding groups like birds and most insects, is woefully inadequate as a concept of specificity." We must take into account the fact that the absence of dioeciousness is a derived exceptional condition, and that those groups in which it is absent are always only subgroups within more inclusive groups of far higher rank in which dioeciousness is the rule. In the light of this fact the question of the species concept in organisms without bisexual reproduction is no more than a relatively subordinate special problem of systematics.

There are further considerations. It is true that in groups lacking bisexual reproduction, such as the parasitic flatworms and certain parthenogenetic groups of animals, the structural differences between ontogenetic, tokogenetic, and phylogenetic relationships (Fig. 6) are blurred. They are not completely missing, however. No one would deny that there is a difference between the ontogenetic relationships that interconnect semaphoronts and the tokogenetic relationships between different individuals even in organisms that lack gonochorism. We characterized phylogenetic relationships as segments in the stream of tokogenetic relationships (p. 20). These segments are also present in groups of organisms without gonochorism. Consequently, even in these groups it is possible to delimit in the fabric of hologenetic relationships an area that lies between the more or less unequivocally phylogenetic relationships on the one hand and the ontogenetic relationships on the other. This area naturally corresponds to the species category of organisms with bisexual reproduction. In groups of organisms without gonochorism the species concept lacks a few criteria that are essential to the concept in general, but it is not so different as might appear at first sight. We will return to this matter in discussing group categories of higher rank.

The practical and theoretical usefulness of the definition of the species as the totality of all individuals that form a reproductive community encounters other difficulties in addition to those in groups without gonochorism. Certain spatial requirements must also be met—a group of individuals genetically capable of producing fertile offspring must also form a reproductive community in fact. These spatial requirements are not always met.

If the territory occupied by a species is divided up into a mosaic of small areas that are more or less completely isolated by barriers to distribution, then instead of a uniform gradient of relations there will be a pattern of smaller reproductive communities corresponding to the mosaic of small areas. The members of these smaller communities will interbreed only rarely, even though the potential capability for interbreeding remains unimpaired. The extraordinarily advanced mathematical treatment of these questions shows, however, that the accidental loss of alleles that accompanies the breaking up of a species into small reproductive communities (Wright's formula) involves a differentiation and, with the correlative interconnection of all gestalt characters, probably a restriction of the capacity to interbreed.

Between the different communities of reproduction separated by barriers to dispersal there will now come about a gradient of relationships similar to that assumed for the individuals in uninterrupted space, with the one difference that the intensity of the gradient depends not only on distance and capacity for migrating but also on the effectiveness of the barriers that subdivide the space. The effectiveness of the same barrier (e.g., a large river or a mountain range) may, of course, differ greatly from species to species.

A graphic presentation of the effectiveness of a barrier to dispersal on the tokogenetic relationships of different close communities of reproduction is shown in Fig. 10. That the species involved is man and the barrier a political boundary has no theoretical significance.

In order to accommodate these facts we could consider including the capability of interbreeding—that is, the possibility of unlimited fertile crosses between all individuals—rather than actual interbreeding, in the species definition. This has in fact been done (the expression "potential community of reproduction," used above), and crossing experiments have long been regarded as one of the most important aids in deciding as to the conspecificity of organisms that differ in form. The importance of this criterion is greatly limited by the fact, already mentioned in the preceding chapter, that the capacity to interbreed may be greatly reduced between different individuals of a species, and that no sharp line can be drawn between unrestricted and completely absent crossability. This is true even for individuals living in the same area, and to an increased degree for individuals from distant parts of the range of the species. Experience shows that there is often a gradient of capability for successful interbreeding that corresponds to the spatial gradient of actual reproductive relationships. According to Pictet (1939), for example, the fertility and vitality of the offspring in butter-

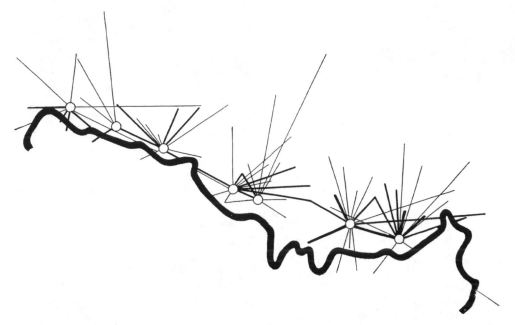

Figure 10. Political boundary as a barrier to marriage. An example of the effect of spatial barriers on reproductive relations between individuals of a widespread species. Circles indicate villages on the boundary, lines the directions and distances of intermarriages with persons from other villages. (Redrawn from Schwidetzki.)

flies decreases as the geographic distance between the parents increases. This gradient may be so strong that successful interbreeding becomes completely impossible between individuals farthest separated from each other or between individuals belonging to narrower reproductive communities. Further examples are given by Dobzhansky and Streisinger (1944), Moore (1949), and Callan and Spurway (1951).

Chorological Relationships of Individuals and Their Significance for the Taxonomy of Lower Group Categories

In the preceding sections the species was defined in simplified form as a reproductive community. This definition required a certain limitation, since the reproductive relations between the members of a species are in part only potential —we cannot assume, particularly in species with large populations, that all individuals of opposite sex actually do interbreed. Above all, the fact that the individuals of a species do not live directly side by side, but are distributed more or less widely in space, was completely disregarded. Because individuals are distributed in space, a factor of the utmost importance in bringing about diversity

—i.e., increasing, or at least altering, diversity—was left out. This factor is of great importance for biological, and above all for phylogenetic systematics.

The significance of space in phylogenetic systematics lies not only in the relations between the members of a species living in it and the other factors of the environment that affect their lives, but also in the reactions that its extent alone has on the relations between the individuals of a species living in it. It is not the relations between the species and the space and its living conditions that are in themselves of primary interest from the systematic standpoint, but the interrelations between the individuals that are distributed throughout the space.

The indisputable fact that each individual requires a certain area in order to live—differing greatly in size for the various ecological types—is one of the most important factors that at least hinders interbreeding between all members of a species. Obviously under otherwise similar conditions neighboring individuals will interbreed oftener than individuals living apart. Consequently, in the sexual relations between each individual and others of opposite sex, there is a gradient whose intensity varies directly with distance and with the active and passive capacities for migration in the different species. Sometimes this is also expressed in gradients of morphological and other characters. The "cline" concept (Huxley 1939, 1940) was created for such gradients in the expression of characters. Rensch (1929) recorded limitation of fertility between distant populations in *Phasianus colchicus* and *Streptopelia chinensis*, and in his other works there are additional examples of gradual transitions from uniform species to species that are split up into numerous narrower communities of reproduction.

From these examples it is evident that even the broadened definition of the species as a community of all individuals between which fruitful interbreeding is possible (instead of actually existing) cannot completely satisfy the natural relationships. It is unacceptable because the end members of such chains or circular complexes of reproductive communities as in the examples above would be considered as not belonging to the same species only because the capacity for reproductive relations between them is completely absent. It would then be necessary to make a completely arbitrary cut between two neighboring members of the total complex. Consequently it is absolutely necessary to include the chorological relations in the definition of the species.

The species would therefore be defined as a complex of spatially distributed reproductive communities, or if we call this relationship in space "vicariance," as a complex of vicarying communities of reproduction. The possibility that all members of a species actually do form a closed (even though only potential) community of reproduction—as is conceivable for a species consisting of a very few individuals living in a very small area—would be the limiting case within this definition.

The word "space" in the foregoing referred primarily to two- or three-dimensional geographic space, but it is necessary to extend the concept of space to the multidimensional environment. This environment is not an entity independent of and in addition to geographic space. On the contrary, it merely characterizes further dimensions of geographic space, corresponding to the multiplicity of living conditions that geographic space offers to the organisms living in it. The relations described above as existing between the species and geographic space also exist in a very similar way between the species and its environment. "In the environment too a portion of space that is occupied by one organism or group of organisms cannot be occupied in the same manner by another" (Dingler 1929). But this too is correct only if the environment is considered in connection with geographic space, because if one speaks of "the occupation of a portion of the environment by organisms," this is only a different—and for many purposes more satisfactory—way of expressing the fact that these organisms have certain habits, or require very definite living conditions in the part of the geographic space that they inhabit. In this sense it is entirely possible for two species or two individuals to have the same mode of life in two different geographic areas, i.e., to have the same coordinates in all dimensions of the environment. Nevertheless they do not occupy the same position "in space" and thus can exist side by side, because their "geographic coordinates" are different. In the following when we speak of a distribution of organisms in the environment, we will completely disregard the geographic dimension and express only the fact that different organisms or species almost without exception have different habits and to this extent can be regarded as distributed in different areas of the environment. For this "distribution in multidimensional space, namely in the environment, this assumption of a place or occupation of an area" Dingler (1929) chose the term "location," and distinguished between "purely geometric, trophic, and temporal location." Under "purely geometric location," for example, Dingler understands the phenomenon that among closely related woodboring beetles one species may occupy the lower, thicker portion of the trunk, another the upper, thinner portion; or that one species constructs the parent galleries in the direction of the trunk axis and the larval galleries at a right angle to the axis, whereas another species proceeds in the opposite manner (large and small pine beetles). The examples given by Eichler (1940) and Szidat (1956) of "topographic specialization" in parasites (e.g., head and body lice, trematodes in different organs), and the different biting habits of bloodsucking insects (e.g., leg bites, in contrast to species that prefer the neck or face) also belong in this category. Hering (1955) named two species of the boring fly *Ceriocera ceratocera* that replace one another "topographically": the nominate form lives in the flower heads, the form *microceras* in the stem marrow of the same plant (*Centauria scabiosa*). The significance of the other "locations" distinguished by Dingler are self-explanatory. No sharp boundary can be drawn between them.

For the systematics of the lower taxa, exactly as in the case of geographic space, it is not so much the relations between the species and its environment or the peculiarities of the individual dimensions or parts of the environment that are important, but rather the reactions on the reproductive relations between the individuals of a species that arise from the diversity of the living conditions—corresponding to the breadth and the obstacles of geographic space. Just as there are widely and less widely distributed species in geographic space, there are species that—figuratively speaking—are restricted to a narrow area of the environment (stenoekous species), and others that colonize vast areas of it (euryoekous species). Although relatively little is known about it, there is probably a gradient of tokogenetic relations between the individuals of euryoekous species similar to the gradient between individuals in species that are widely distributed geographically. Thus we may suspect that there is more frequent interbreeding between individuals of a species occupying the same environmental segment than between individuals from different segments.

It is known, for example, that the larvae of the agromyzid fly *Phytomyza atricornis* can develop in more than 300 species of host plants. Even without detailed investigation we can assume that, at least in places where one or another of the host plants forms large stands, the individuals living in the host plant will mate with one another oftener than with individuals living in other host plants. In this case we must assume, in the absence of evidence to the contrary, that the gradient in the reproductive relations between individuals whose larvae came from different host plants concerns only the relative frequency of matings, and that the capacity for successful reproduction between all members of the species is unimpaired. Moreover, in this example the effects on reproductive relations resulting from species differences in the host plants, which (at least where these form large stands) result from geographic differences in location that in turn reflect differences in the environment, cannot be distinguished sharply from the effects that geographic space alone exerts on the reproductive relations of the insects. This distinction is more easily made in other cases. The larvae of the North American fruitfly *Rhagoletis pomonella* live in both apples and blueberries. In North America both host plants often have the same station, and the adaptation of the fly to these two very different host plants is probably to be thought of as having arisen through the frequent intimate spatial proximity of the two. Imagos reared from the two host plants are morphologically indistinguishable. Whether fertile crossing between them is possible, and the extent to which such crosses occur in nature, are unknown. Since larvae from blueberries can rarely be brought to developmental completion in apples and vice versa (Lathrop and Nickels 1931), it must be assumed—even though there are other possible explanations—that at least the actual and perhaps also the potential reproductive relations between imagos from the two host plants are more or less restricted. These surmises were later confirmed by the investigations of Pickett and Neary (1940).

How adaptations to such different host plants can arise is indicated by the observations of Thorpe (1939). Imagos of *Drosophila* are normally repelled by the odor of peppermint, but if larvae are reared in media with 0.5% peppermint essence the resulting imagos are perceptibly attracted by peppermint. Thus it is easy to imagine the adaptation of *Rhagoletis pomonella* to blueberries as having arisen by accidental transfer of the young larvae to blueberries (e.g., by falling from an apple tree); by completing their development there, a change in the olfactory reactions of the imagos was brought about. Hering (1926) gives a number of examples of accidental colonization of abnormal host plants that could lead to ecological vicariance in this way. In all these examples nothing is known of the actual or potential reproductive relations, and partial restriction can only be regarded as very probable. A number of cases are known—under the designations "biological or physiological species," "ecological races or subspecies," "species sorores" (Schröter), "habit races" (Magnus), "sibling species," "espèces jumelles" (Cuénot), "dual species" (Pryer 1886)—in which vicariance in the environment has been shown to be accompanied by restriction or complete impossibility of crossing. Numerous examples of such biological species, habit species, or whatever they may be called, have been demonstrated in insects that are parasitic on plants and animals. We will only point out that it has been recognized to an increasing degree that species formerly regarded as very polyphagous are in many cases complexes of smaller subdivisions, each of which is closely tied to a particular sector of the circle of host plants of the species as a whole. (See also Hering's "Minenstudien," and Barnes 1953.)

Under the name "Hopkins' rule of choice" it is well known to ecologists that in polyphagous insects (and in nematodes) the females often choose for their offspring the same species of host plant on which they themselves developed. Unfortunately few systematists are aware of this rule, as I know from experience, and consequently pay too little attention to it in their works. A detailed presentation is given by Friederichs (1930).

Such vicariances are not limited to the "trophic dimension" of the environment. Hering (1936) described two ecological races of the species *Noeeta pupillata* that cannot be distinguished for certain in the imaginal stage. One of these overwinters in the pupal stage and has a motile puparium (*N. p. pupillata*), while the other, with a fixed puparium, overwinters in the larval stage (*N. p. pardalina*). This could be called vicariance in the temporal dimension of the environment. It should not be difficult to find examples of vicariance for any dimension of the environment (cf. the various categories of subspecies distinguished by Edwards 1954), although the distinctions between the various dimensions are more or less arbitrary and not wholly satisfactory.

In the foregoing presentation of the vicariance relations in the environment insofar as they relate to the species category, we often had to leave unclear in our examples whether closely related species or complexes of ecological ("biological") races within a species were involved. This is because in contemporary systematics the decision as to whether obviously closely related but sexually more or less isolated groups—insofar as they do not obviously replace one another geographically—are regarded as different species or merely as ecological "races" and the like often depends on whether they can be distinguished morphologically to some degree. For practical reasons I propose general acceptance of the following statement: "species, genera, and higher systematic groups must not be based on exclusively biological differences" (Plate 1914). The uninterrupted relationship between geographic and ecological-physiological variability that was pointed out above shows that we can achieve a correct formulation of the species concept, and a fixing of the species limits that approaches the actual situation in nature, only if we start from the fact that there is vicariance in the dimensions of the environment as well as in geographic space. Each of these two forms of vicariance is to be understood only as an extreme case in a host of possibilities in which all intermediate possibilities between the two extremes are realized. Consequently it is inadmissible in taxonomy to treat vicariance in the geographic dimensions of space differently than vicariance in the ecological and physiological dimensions.

Biological species are currently a favorite subject of evolutionary research. The question at issue is whether geographic barriers must be involved in the splitting of a species leading to the origin of new species (see Fig. 4), or whether—according to a popular formulation—a "sympatric" origin of species is not possible in addition to the "allopatric."

Mayr (1957) has discussed the "conceivable forms of speciation." According to him the numerous hypotheses that have been advanced fall into twelve ideally possible methods of speciation. "In reality, however, only polyploidy and geographic speciation need be considered, in the animal kingdom almost exclusively the latter." "It would be completely false to say that species can arise either through geographic isolation or by the development of isolating mechanisms. Such an alternative is nonsense. Species can maintain themselves only if they have genetic isolating mechanisms. The indispensability of geographic isolation consists in the fact that it makes possible the harmonious genetic reconstruction that is necessary for the origin of new species-specific isolating mechanisms." The so-called "ecological" origin of species is postulated, according to Mayr, "with the regularity of a clock almost every year by some author or other. But for almost a hundred years it has always been invalidated with thorough proofs." In his further discussion Mayr considerably restricts these somewhat apodictic formulations: "The expression 'geographic isolation' should not be taken too literally. 'Spatial separation' would probably be a more accurate expression, because the only thing important is the undisturbedness of the separated gene reservoir. In parasites, for example, under certain circumstances the gene flow may be cut off if a parasite founds a colony on a new host. Such a spatial separation is biologically completely equivalent to the geographic separation of free-living species."

It seems to me that in this last remark Mayr loses sight of what he set out to prove. If the distinction between geographic and ecological isolation or allopatric and sympatric isolation is to make any sense, "geographic" isolation of a population can only mean separation by barriers whose material existence hinders meeting sexual partners from another population to such an extent that individuals never or only very rarely physically overcome the barriers—at any rate too rarely to guarantee a free interchange of genes. But this is not the case when "a parasite founds a colony on a new host," or if it succeeds (see above) in colonizing a new organ of the same host. If one tries to show that only geographic isolation can make possible the stabilizing of a new gene pool, it is not important whether a certain kind of "spatial separation" is "completely equivalent biologically to geographic separation," but rather whether it is geographic separation in the specified sense or not.

For practical systematics the question of whether there is sympatric speciation in addition to allopatric speciation is not decisive. Of greater importance is the fact that it is extremely difficult, even practically impossible, to decide whether a group of individuals characterized by particular morphological characters as a closed reproductive community is really a closed reproductive community, a complex vicarying but not completely isolated reproductive community, or a group of species that is completely isolated genetically. This is readily explainable and merely the visible expression of the fact that, with Dobzhansky (see Ulrich 1942), we must regard the species as "the study of the evolutionary process in which groups of forms that hitherto have interbred, or at least were capable of interbreeding, become divided into two or more separate groups that for physical reasons cannot interbreed." Perhaps it would be even more appropriate to characterize the species as the condition of equilibrium that in nature tends to come about between the pressure toward diversity on the one hand, and the conservative principle that is expressed in the bisexual mode of reproduction on the other. Systematic groups in such a condition of equilibrium are rarely found in nature, but we cannot conclude from this that the species is an artificial invention, and that there are no generally valid principles, devoid of subjective arbitrariness, by which they can be delimited. To do so would be exactly as reasonable as to deny that organisms are constructed of cells because on a section through an organism we might find fewer typical cells than cells in all phases of division. The cell too can be interpreted as a condition of equilibrium that living substance tends to achieve at a certain level. The example of the cell also shows clearly that if the number of cells in a section is to be counted, what is counted as one and what as two cells is to a considerable extent left to the inclination of the observer. This practical uncertainty would not in the least argue against the real existence of cells, or against the theoretically completely unequivocal definiteness of the content and extent of the cell concept. In this sense, but only in this sense, we can agree that it is "a matter of convention" where the line between species is drawn (Zimmermann 1931), and that "what we call a species is somehow an artificial thing, liable to subjective interpretation and variation in the course of time" (Dingler 1929). It is probably better to avoid such mistaken formulations altogether.

It might seem that the difficulties surrounding the species concept, or the impossibility of determining the species limits with certainty in some cases, would destroy or at least considerably disturb the intent of phylogenetic systematics to subsume in groups (taxa) of higher rank those species that arose by the split-ting up of a common stem species. This, however, is not the case.

For the taxonomy of the higher categories it is without essential importance that the process of speciation is usually a relatively long, drawn out process, so that a cross section through the species stock at a particular time—the present, for example—shows the several species in different phases of their cleavage. The subsuming of species in group categories of higher order according to the degree of their phylogenetic relationship depends on a process of measuring—measuring the time since the fundamental event of speciation, i.e., the species cleavage that led to the origin of the higher taxon. Since this involves time spans of geological magnitude, which are the longer the higher the rank of the taxon, the time re-quired for a phylogenetically significant process of speciation is negligible. Sys-tematics can neglect it without being charged with superficiality and without

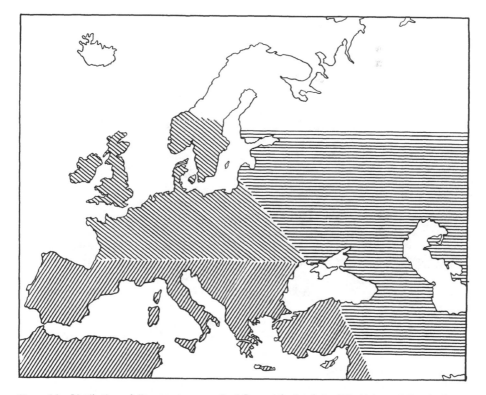

Figure 11. Distribution of "temperature races" of **Drosophila funebris.** SW: higher relative vitality at 29° C, lower at 15°; NW: lower relative vitality at 29°, higher at 15°; E: higher vitality than the western races at both 29° and 15°. (Redrawn from Timoféeff.)

fear of failing in its aim. Involved here is an essential factor of any process of measuring: "the adaptation of the unit of measurement to the magnitude of the distance to be measured" (Kienle 1948).

Insofar as phylogenetics (phylogenetic systematics) is engaged in constructing groups that, measured against the total course of phylogenetic development, appear as close sequences of phases in a process of speciation, it has the right and the obligation to use very exact measuring devices, and to make subtle distinctions between the phases. In other areas of its work it can disregard these without impairing the exactness of its results, and base its arguments on the immutable nucleus of the concept of the species and speciation: community of reproduction and disintegration of a community of reproduction into genetically isolated successor communities of reproduction.

Distinguishability of the different subspecies and other subunits of a species (whatever they are called in particular cases) is tied to the presence of morphological differences. The nature of the holomorphological differences may vary a great deal. Like the individual variants, they may involve differences in all the various dimensions of the holomorph—morphological, the most diverse forms of physiological, ethological, psychological, and other features—and these may be combined in the most manifold ways. Numerous examples can be found in the works of Rensch (see also Figs. 11 and 12). The different forms of allomorphism can be distinguished from one another and from the characters of the species and higher taxa, at least within limited areas of applicability, and with suitable experience. There are, however, practically no rules for differentiating subspecies. First of all, the extent of the differences between subspecies varies to such a degree that it is almost completely worthless in deciding whether two forms that differ in a particular way must be considered subspecies of a single species or as members of different species or even genera.

Already in 1866 the botanist Anton Kerner von Marilaun, in a paper on "good and bad species," recognized that "the degree of divergence is unimportant," and Kleinschmidt (1940) states categorically that "formenkreis differences may be subtle, race differences need not be subtle." These statements and all similar ones relate to the geographic doctrine of rassenkreisen, and from this standpoint many groups are called different species that from our standpoint—which emphasizes ecological vicariance more strongly—would be regarded as subspecies of a species composed of several ecologically and geographically vicarying components. In any case the correlation between degree of holomorphic similarity and phylogenetic relationship is far too loose to be of much value in distinguishing between the category steps of the subspecies and species.

The situation is not much better with regard to the quality of the differences. Simple differences in proportion, differences in the lengths of body parts that can be reduced to the laws of allometric growth, size differences, and definite color differences commonly occur as subspecific characters. None of these characters and differences by themselves ever constitute proof; their presence merely justi-

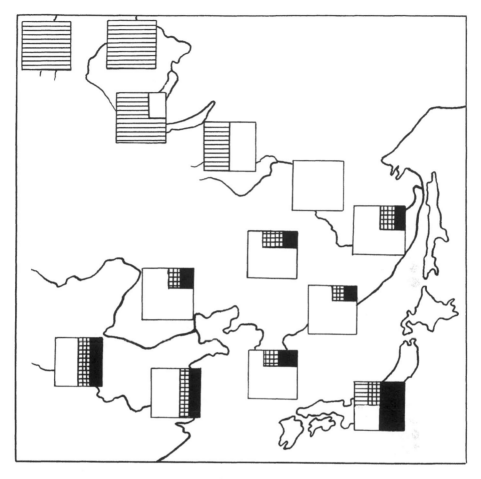

Figure 12. Geographic distribution of four color patterns of the beetle **Harmonia axyridis** in eastern Asia. White, var. **signata**; horizontal lines, var. **axyridis**; crosshatched, var. **spectabilis**; black, var. **conspicua**. (After Dobzhansky.)

fies the suspicion that subspecific differences are present. Whether the differences are associated with vicariance in space is always decisive. Biological systematics has relied predominantly on morphological characters in uncovering the subspecific differentiation of a species solely because morphological differences are particularly easy to recognize. It should not be forgotten, however, that this produces a one-sided picture that does not represent the true relationships entirely correctly. The ultimate aim of taxonomy at the species level must be to recognize and distinguish as subspecies all vicarying reproductive communities, at least insofar as (in the case of purely geographic vicariance) these are not merely externally limited but also have definite distinguishing characters. Such distinguishing characters need not be purely morphological. It is entirely possible for vicarying subspecies to differ only statistically, with certain variants (allomorphisms) appearing in different frequencies in two spatially separated reproductive communities (Fig. 12). Ideally such reproductive communities must also be distinguished from one another as subspecies, even though it may be practically impossible to say which subspecies individuals belong to if their spatial origin is unknown, and even though certain individuals (variants) of one subspecies may be structurally more similar to certain individuals of the other subspecies than to members of their own subspecies.

Since the individual peculiarities of the holomorph need not differ to the same degree among the individual vicariants, consideration of only one character or one group of characters will produce an erroneous picture of the actual situation in nature. Fig. 13 is a greatly simplified representation—in which a multidimensional structure is projected in simplified form onto a two-dimensional plane—intended to show how the results of subspecific differentiation may differ if single characters of the holomorphic gestalt are made the basis for separation, and that none of these results need correspond to the actual genetic structure of the species (Fig. 13e, f).

Many of these questions have been actively discussed in the journal *Systematic Zoology*, without producing essentially new viewpoints. Christiansen (1958) and Hagmeier (1958) have presented examples showing that the delimitation of a "subspecies" varies with the characters used. This makes the old-style rassenkreis taxonomy, which tried to divide up the species into "subspecies" according to more or less easily determinable external characters, somewhat questionable. The actual composition of the species out of numerous part populations is only incompletely expressed in this way. On the other hand, there is little inclination today to accord the smallest recognizable part population of a species the rank of "subspecies" (see Rogers 1954). Gilmour and Gregor (see Edwards 1954) have proposed the term "demes" for these units. "The number of different demes recognized by taxonomists depends upon the thoroughness with which the species including them has been investigated" (Edwards 1954). The question of terminology is without particular significance for us. It is decisive that even today what we said above concerning the broadening of the definition of the species by including the criterion of vicariance can only be confirmed.

The Species Category in the Time Dimension.
The Species Concept and Paleontology

Discussion of the questions arising from the distribution of species in space has already led us to some extent beyond the species category into the taxonomy of the higher group categories. The impossibility of discussing the species concept apart from the problems of the taxonomy of the higher taxa is even more sharply

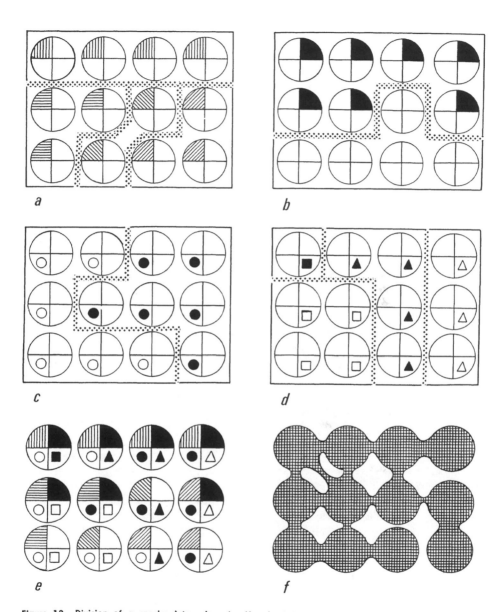

Figure 13. Division of a species into subspecies if only (a) morphological, (b) physiological, (c) psychological, or (d) ethological differences are observed. Complete separation of a species into subspecies according to all "holomorphological" indexes would—corresponding to the actual presence of vicarying closely related reproductive communities—produce the picture shown in e. The intensity of reproductive relationships between the individuals of different subspecies is represented by f.

evident when the question of the duration of species in time is raised. How should systematics deal with the classification of fossil organisms in a system based on recent, genetically defined species in the temporal cross section that the existing fauna represents, and which tries to represent in its higher taxa the phylogenetic relationships between these species?

It would be most appropriate to discuss this whole complex of questions later in connection with the tasks and method of systematics in the area of the higher taxa. But the special question of "the species concept and paleontology" has lately received special attention, and has even been the subject of a symposium. Consequently it seems advisable to discuss it here in connection with the preceding discussion of the species category.

As a starting point for evaluating this complex of questions we must go back to our derivation of the concept of phylogenetic relationships from the consideration of the fabric of hologenetic relationships that interconnect all "semaphoronts." We concluded that the structure of the phylogenetic relationships can be represented accurately by a diagram (Fig. 4). This, together with the fact that our diagram corresponds to the definition of a hierarchy in the sense of Woodger and Gregg, supplied the premise for our conclusion that a hierarchic system is the appropriate representational form of phylogenetic systematics.

Our diagram (Fig. 4) uses two different symbols: circles representing species, and arrows representing relationships between these species. This use of symbols determines the form in which "time" appears in our diagram: the circles symbolize periods of time, the arrows represent points in time in the history of the species. This is directly evident from a comparison with the more detailed presentation of the "stream of tokogenetic relationships" (Fig. 14). The limits of the species in a longitudinal section through time would consequently be determined by two processes of speciation: the one through which it arose as an independent reproductive community, and the other through which the descendants of this initial population ceased to exist as a homogeneous reproductive community.

Objections have been raised to such a delimitation of the species by two successive processes of speciation. Most important for the theory of systematics is the objection that it singles out and unduly emphasizes the origin of species by cleavage. An equally important process of speciation is transformation, which is represented in Fig. 15. In this diagram all points in time $(t_1\text{-}t_9)$ that are important in the history of a monophyletic group of species are shown.

In our initial diagram of the process of speciation (Fig. 4) we assumed a splitting of the initial population into two populations of almost equal size and, after the development of the genetic isolating mechanism from the initial population, almost equally strongly different successor populations. Such a process would present no theoretical or practical difficulties for the determination of species limits after cleavage had taken place, but it represents only one possibility (speciation by division, see Brown 1958). Perhaps more commonly only a small partial population splits off from the parent population and becomes a new species

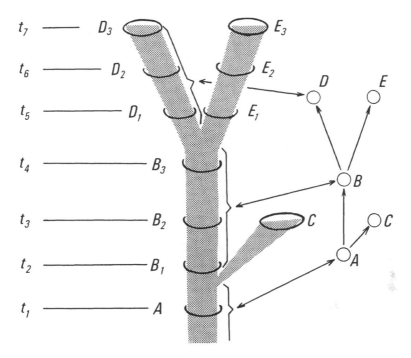

Figure 14. The species category in the time dimension. Explanation in text.

(speciation by colonization, see Figs. 16 and 17 and the ideas of genetics on the dynamics of area boundaries). In such cases it can be assumed with certainty that only the species arising from the original small split-off population will be notably different from the parent population.

We can probably include in this mode of speciation all those cases commonly encountered in insects in which, in addition to a common widely distributed euryoekous species, there is a second species that is very little different, less widely distributed, and stenoekous. Herting (1960) cites a number of such cases in the Tachinidae (Diptera). *Blondelia nigripes* is known from 61 species of macrolepidoptera, 7 species of microlepidoptera, and 8 species of sawflies. The evidently closely related *Blondelia piniarius*, on the contrary, is a specific parasite of *Bupalus piniarius*. Herting assumes that "*B. nigripes* is the stem species from which *B. piniarius* arose by cleavage." Conditions are similar in *Phryxe vulgaris* (a polyphagous parasite of the caterpillars of macrolepidoptera, noctiids, and geometrids) and *Phryxe erythrostoma* (a specific parasite of *Hyloicus pinastri*), and in *Myxexoristops blondeli* (in various species of sawflies) and *M. hertingi* (specializes on *Acantholyda erythrocephala*).

Indications at least of this kind of speciation can occasionally be observed even

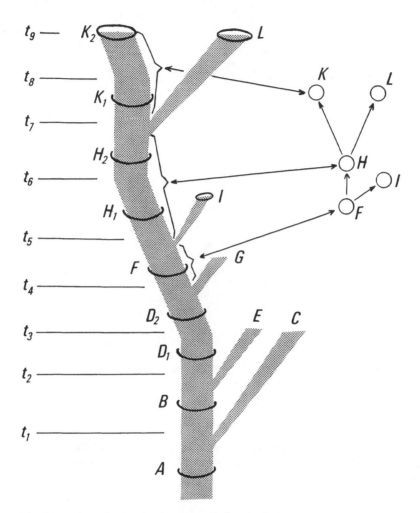

Figure 15. The species concept and paleontology. Explanation in text.

in the present. Barnes (1953) reports that the gall wasp *Stenodiplosis geniculati* Reuter attacks both *Alopecurus* and *Dactylis* in Europe. The populations from these two grasses are morphologically indistinguishable. *S. geniculati* has been introduced into New Zealand from Europe. There it also attacks both grasses, but the populations from *Dactylis* can be distinguished by minute but distinct antennal characters from the populations living on *Alopecurus*. Consequently they have been taxonomically separated from the main form as "var. *dactylidis*."

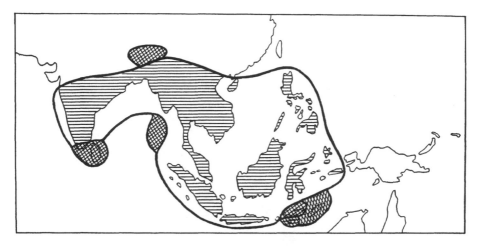

Figure 16. Dynamics of areal boundaries. Distribution of the tree snake **Dendrophis pictus.** Subspecia-
tion at the margins of the total area. (Redrawn from Meise and Hennig.)

"It was suggested that in New Zealand *S. geniculati* changed over from Foxtail
to Coxfoot some years ago, and in the course of twenty years or so had developed
the morphological differentiation already mentioned." Barnes assumes that in
Europe (England and Ireland) the shift from *Alopecurus* to *Dactylis* took place
only recently, "and consequently there had not been sufficient time for morphological characters to have evolved. The alternative is that they were not firmly
established and isolated on the new host plant." Unfortunately nothing is known
regarding the genetic isolation of the populations.

If we adhere to our original definition, according to which "one and the same"
species is determined by two processes of speciation, and assume that in New
Zealand the form "dactylidis" (with strict application of the genetic concept of
the species) actually represents a distinct species, then we must say that the
species described by Reuter 1895 under the name "geniculati" was a different
species from the one we characterize by that name today. This would be true
even though the living generation differs in no way from the one that lived in
Reuter's time, and even though the two generations would doubtless form a
homogeneous reproductive community if they could be brought together. Such
a statement appears paradoxical to the logical human mind, and obviously no
systematist would be prepared, on the basis of such reflections, to give another
name to the species that is still called *Stenodiplosis geniculati*.

Paleontology faces similar problems. Let us take, for example, the portion of
the hypothetical phylogenetic tree shown in Fig. 15. According to the ideas so
far developed we would have to assume two different species, B and C, between
the points in time t_1 and t_2, both of which originated from the common stem

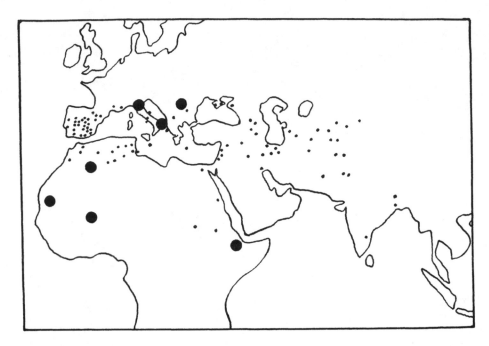

Figure 17. Dynamics of areal boundaries. Distribution of the fly **Phlebotomus papatasii.** The large black circles are the localities ("terra typica") of individuals described as representing species of their own. Almost all lie at the margin of the range of the species.

species "A." Paleontology, to which individuals from both populations A and B might be known, would be unable to conclude that these individuals belong to "different species." Thus here during a process of speciation (to the point in time t_1) one of the two "daughter species" has not changed relative to the stem species.

On the other hand there is the possibility that the descendants of a population may change without a cleavage having taken place. For example paleontology, to which individuals from populations D_1 and D_2 are known, could not determine on the basis of the ideas we have developed so far whether it was dealing with "one and the same" species that, at the point in time T_3, had merely changed without splitting up.

Apparently there are two ways of escaping this dilemma. One is to incorporate in the genetical species concept a morphological principle criterion for the time dimension—in the final analysis, therefore, for use in paleontology—leaving the task of determining species boundaries by species cleavage processes to the principle we have been defending. This course has been followed in the proposal of the concepts of "paleospecies" and "chronospecies," which are purely morphologically defined concepts. In Fig. 15 the populations A, B, and C, and H_2 and I would belong to a paleospecies or chronospecies. On the other hand, the populations D_1 and D_2, and H_1 and H_2, would belong to different paleospecies. These concepts also circumvent the difficulties that only dead individuals (more accurately, semaphoronts) are available to paleontology, and in the most favorable cases only pieces of the total fabric of characters of these semaphoronts. For this reason paleontology is actually never in a position to determine whether corresponding or different semaphoronts are members of the same or of different reproductive communities, a possibility that always exists at least in principle for neozoology.

But is anything really gained from the purely morphological concepts of paleospecies and chronospecies, or from including a morphological principal criterion in the genetic species concept of neozoology in order to establish species limits in the time dimension? In paleontology the incompleteness of the fossil record dictates the use of purely morphological species concepts as a basis for a purely formal classification. So long as only a few semaphoronts are known it would create no difficulties if, for example, the species A, B, D_1 or H_1, and H_2-I are distinguished. But if more and more individuals from the sequence of generations in these populations became known, a purely morphological distinction of the species would eventually become impossible. "As the fossil record becomes more completely known, the problems are not therefore likely to be resolved, as is generally supposed, but will become more acute" (Clark 1956). Consequently we can agree with Cain (1956): "the paleospecies is the result of an unhappy attempt to impose a taxonomy of discontinuous groups on a continuous series."

Thus we come to the second way of establishing species limits in the time dimension. This is the possibility from which we originally started: delimiting the species by two successive processes of speciation. Admittedly this does not eliminate the practical difficulties of paleontology. For the practical determination of species limits all processes of speciation must be known: in addition to those shown in Fig. 15, we would have to know all those that led to the origin of species that became extinct without leaving successors. Since all species cleavages will practically never be known, it will never be possible to say whether fossil individuals from different time horizons belong to the same or to different species. We have seen that this would not be possible even by using a purely morphological species concept.

There are certain conceptual difficulties in the case of species delimitation by two successive processes of speciation. Simpson (1951) says it is "not useful taxonomy to classify as the same thing throughout" a species that has changed between two processes of speciation (for example, the species H_1-H_2 in Fig. 15). In opposition to this we can point to the phenomenon of metamorphosis; different metamorphic stages of the same individual are not "one and the same thing" in this sense either. It is not clear where the difference is supposed to lie when we speak analogously of transformation stages of one and the same species.

The idea that under certain circumstances populations (e.g., A, B, D₁ in Fig. 15) are to be regarded as "different species" although they are not at all "different" morphologically presents great difficulties to many authors (Alberti 1955, 1957). A narrowness of outlook in the logical interpretation of phylogenetic systematics is expressed in discussions of this question. This same narrowness is often encountered in other connections. In paleontology it leads to considering technical difficulties prematurely in purely theoretical discussions, and using comparisons from the life history and mode of reproduction of individuals for interpreting the history of species beyond the point to which such comparisons are fruitful.

We must hold unswervingly to the fact that the starting point of all our insights and efforts is the geological present. The genetic species concept, which in turn rests on the concepts of "population," "reproductive community," and "genealogical fabric of relationships among semaphoronts," is the result of investigations of phenomena of the present. Of course, this present condition, for which the grouping of organisms into species is characteristic, is the result of evolutionary processes, among which the processes of speciation assume a preferred place. The diversity of existing discrete species arose by processes of speciation. To the extent that purely morphological transformations are involved, the phylogenetic relations of living species are of essential importance to a detailed understanding of evolutionary processes. We measure the degree of these relationships according to the temporal sequence of the processes of speciation that led to the origin of the living species. For us the phylogenetic system is a medium for presenting the phylogenetic relationships of species (primarily the living species) as accurately and clearly as possible. Such a picture of phylogenetic relationships can be a system of hierarchic type only if in its plan of construction the species is regarded as the unit that undergoes division. This is possible only if two successive processes of species cleavage are assumed to be the temporal delimitation of the existence of a species. "Stem species," from which two or more recent species have arisen (by whatever type of speciation), do not occur in the hierarchic system of recent species. They can be provided only by paleontology.

We know that paleontology is not able, for a particular horizon in time, to determine species limits with the same exactness as neozoology. Paleontology is dependent exclusively on morphological criteria, and even with these it can work only considerably more coarsely than is possible with recent species. This does not mean that we must give up and falsify our goal of constructing a phylogenetic system of recent species because of the limitations of paleontology, but merely that our expectations as to the extent to which paleontology can help in solving our problems are limited.

Even the "species" of paleontology, which are necessarily determined by exclusively morphological criteria, are not qualitatively different from what we must understand as natural "species" on the basis of irrefutable considerations—the sequence of generations of reproductive communities whose boundaries in the time dimension are determined by processes of cleavage. If the concepts of "paleospecies" and "chronospecies" are to have any meaning at all, it can only be that they point forcefully to the fact that what paleontology simply calls "species" are often not the true species of the phylogenetic system, but rather sections of the family tree that represent more or less coarse approximations to the actual species. Nor does this mean that the "species" of paleontology cannot be included in our phylogenetic system. Under certain circumstances (see below) this is entirely possible. But the limitations inherent in a comparison between paleontological and recent "species" must always be kept clearly in mind, and we must be careful not to falsify our phylogenetic system through paleontology and thus render the system useless for the goal for which it is intended. The following chapters will present repeated opportunities to show in detail how this is to be understood. It is to be emphasized that there are not several (theoretically) different species concepts for neozoology and paleontology; it is merely that the difference in practical possibilities brings about a difference in the exactness with which it is possible to delineate what we must call the natural species. This consideration also contains the requirement to increase the exactness wherever possible until the theoretical limits are reached.

Summary

The semaphoront (the character bearer) must be regarded as the element of systematics because, in a system in which the genetic relationships between different things that succeed one another in time are to be represented, we cannot work with elements that change with time. Accordingly the semaphoront corresponds to the individual in a certain, theoretically infinitely small, time span of its life, during which it can be considered unchangeable. In this sense the individual is to be regarded as the lowest taxonomic group category; it includes those semaphoronts that are connected by genetic relationships. Between individuals there are tokogenetic relationships. With this name we denote the relationships between the individual and its descendents and predecessors of the first degree.

All "character bearers" that ever lived appear—ordered by the ontogenetic and tokogenetic relationships that connect them—in the form of a continuous stream that extends from the beginning of the history of life down to the present. This stream is divided into distinct portions by the phenomenon of sexual reproduction. Sexual reproduction has the effect that contemporaneous individuals appear collected into groups whose boundaries are determined by the fact that individuals of opposite sex can reproduce unrestrictedly only within these groups. At any particular moment every individual is a member of such a group. These reproductive communities we call species. As a result of their existence the stream of semaphoronts and groups of semaphoronts (individuals) is divided up into segments, primarily in the time cross section. In the direction parallel to the time dimension the stream is divided into segments by the occurrence of processes of speciation: at certain times the homogeneous reproductive communities (species) become split up—one species divides into two or more. Thus the temporal duration of a species is determined by two processes of speciation: the one to which it owes its origin as an independent reproductive community, and the one that divides it into two or more reproductive communities.

Since the speciation process—the complete genetic isolation of successor reproductive communities from an original reproductive community—requires a long time, there is a certain unsharpness in the delimitation of species. In the time cross section this is manifested particularly in the existence of all kinds of intermediate stages between species that are completely genetically isolated and others that break up into more or less incompletely isolated reproductive communities. This fact was taken into account by incorporating the criterion of vicariance into the species definition, since the narrower reproductive communities replace one another in space.

The concept of space cannot be restricted to geographical space, but must be extended to the multidimensional environment, which includes geographical space.

Theoretically the broad formulation of the species concept that results from including the criterion of vicariance could be incontestably based on the fact that the limits of accuracy in its applicability are set by the duration of the speciation process. But often speciation is not concluded with the formation of vicarying reproductive communities. Consequently a narrower formulation of the species concept would not increase its usefulness in the taxonomy of the higher taxa, but would only make its use more uncertain. On the other hand, for other purposes it may be desirable to choose a smaller group category (e.g., the population, or a close reproductive community) as a unit.

Theoretically it is always possible to determine whether contemporary individuals belong to the same or to different species, but practically it is extraordinarily difficult and consequently rarely realizable. With fossil organisms it is never possible to determine directly whether individuals belong to the same or to different genetic communities: it is impossible in the time cross section because the possibility of testing whether particular individuals belong to the same or to different reproductive communities is connected with the life of these individuals. The determinability of species boundaries in the time dimension presupposes that they can be determined in the time cross section. Moreover, we would have to know all instances of species cleavage.

Since direct determination of the boundaries of genetic species can never be achieved in paleontology and only rarely in neozoology, we must consider to what extent accessory methods can be used. This means particularly testing the extent to which the boundaries of genetic species can be recognized with the aid of morphological (in neozoology, holomorphological) criteria. The multidimensional gestalt of the semaphoronts was called holomorphy.

We found that while there are definite correlations between the genetic relations of the semaphoronts and their (holo) morphological characters, these always have only limited areas of applicability and are never complete. The limits of applicability must be determined empirically in each case. Comparative holomorphy can be used as an accessory science for recognizing genetic relationships that are to be presented in the taxonomic system.

Thus the "species" recognized with the aid of morphological criteria in neozoology, and especially in paleozoology, correspond only more or less inaccurately with the genetic species. No fundamental contrast between a systematics based on the genetic species concept and one based on practical taxonomy can be constructed from this. It simply makes it possible to determine the exactness with which practical taxonomy can accomplish the task posed by the theory of systematics.

Perhaps an account showing how practical work on the lower taxa looks if it is considered in the light of these results will illuminate the preceding rather theoretical discussion in somewhat more detail. At the beginning is the description of new species. When an author describes a new species he erects a hypothesis, the hypothesis that the specimens he is describing belong to a separate, previously unknown, reproductive community, all of whose members can be recognized with the aid of the characters given in the description. All further work serves to verify, and in some cases to extend, this hypothesis. In principle, verification is possible only by proving that unimpaired reproductive relations are actually possible between all individuals with the specified characters. Such verification is rarely made, because extensive field observation, breeding experiments, crossing experiments, and (with parasites and phytophages) transplantation experiments are required. There are many hundreds of thousands, perhaps even millions, of species of animals, and the necessary investigations on all these species would require centuries of intensive work. Perhaps the systematics of the coming centuries will actually accomplish this task. But it is wrong to reproach the systematist for not rigorously verifying his hypothesis when he describes a new species, or to imply that he is using outmoded morphological species concepts.

Further research may show that the characters used in the original description represent only a part of the range of variation of an actually existing reproductive community. In such a case the "diagnosis" of the species will have to be broadened. It may also happen that, at least with morphological methods—even using modern statistical techniques—no boundary can be drawn between the proposed new species and an already known species. The supposed new species is then dropped; its presumed diagnostic characters are merely a portion of the individual variability of an already known species whose range of variation was inadequately known.

In other cases it will be found that the new species is only a geographic vicariant of a reproductive community already known as a "species." The two "species" will then be regarded as subspecies of a polytypic species, a geographic rassenkreis. In recent decades, particularly among birds and other vertebrates, such gathering together has led to a considerable reduction in the number of "species." On the other hand, the group of individuals with the originally described or subsequently increased diagnostic characters may prove to break up into a number of distinguishable vicarying reproductive communities, which are then given subspecific names.

The use of experimental methods may prove that groups of individuals previously regarded as species are only seasonally dimorphic generations of one and the same species.

The last stage, and in many ways the most difficult to reach, is the insight that a group described as a "species" and characterized by definite diagnostic characters really consists of several closed communities of reproduction that replace one another in the "environment." If it can be shown that these reproductive communities represent species that are truly genetically isolated, one would not hesitate to recognize them even though it is not possible to distinguish such species in the collections. Here no distinction can be made between the systematist as a "museum man," who "wants to arrange his material, i.e., to attach species labels to it and file it under species designations" (Martini 1938), and the modern genetically oriented biologist.

In all cases in which there is no proof of complete (or practically complete) isolation of reproductive communities that are morphologically indistinguishable but which replace one another in the environment, the systematist prefers to speak of polytypic species with various ecological subspecies. He is justified by the fact that the criterion of genetic isolation is usually even more carelessly disregarded in deciding whether reproductive communities that replace one another in *geographic space* are to be regarded as subspecies of a rassenkreis or as separate species. Consequently, in distinguishing ecological as well as geographic rassenkreise, he is guilty of no particular offense against the modern genetic species concept so long as the question of genetic isolation has not been satisfactorily proven.

Because of the great importance of taxonomy to other branches of theoretical and applied biology the systematist will, if circumstances permit, avoid any laxity in using the criteria of the modern genetical species concept, which he might otherwise permit himself in view of the overwhelming abundance of his other tasks. "A pest which is limited to one crop (e.g., a particular grass) is a distinct pest from the grower's point of view and not to be confused with another pest which will attack another crop (grass)" (Barnes 1953). Thus for the farmer and for practical plant protection it is of the utmost importance whether two populations of gall wasps on two different host plants are genetically isolated from one another and therefore restricted to their respective host-plant species. This is the same as asking whether they are different species in the strict sense of the genetic species concept. In this case the systematist will recognize two populations on different host plants as different species, even if they are morphologically indistinguishable, without reference to the criterion of vicariance, if he is certain they are genetically isolated from one another.

Thus delineation of the living genetic species is a long-continued operation. The process has advanced to very different levels in the different groups of animals. In many groups there are still numerous "species" of which no more is known than the morphological characters that were given in the original description. In other groups the process of approaching a complete delineation of the genetic species and their intraspecific make-up is well advanced, because it was possible and necessary to employ the whole intricate apparatus of the most refined experimental and statistical methods. In either case the ultimate objective is the genetic species concept.

The complexity of the relations often faced by systematics in work on the lower taxa may be clarified by an example from the systematics of the mosquitoes. We will follow the account given by Rozeboom and Kitzmiller (1958). *Culex pipiens* was described by Linnaeus as a "species." It "has a palearctic distribution and is also found in certain other areas such as Australia and more elevated regions in East and South Africa. Within this broad range, populations have been designated as subspecies of *pipiens*, but their position is as yet uncertain. These include *pallens* Coquillet of the Oriental region, *australicus* Dobrotworsky and Drummond in Australia, and *berbericus* in North Africa."

Culex fatigans Wiedemann was also described as a separate species. This "species" is often regarded as a geographic subspecies of *Culex pipiens*: "*Culex pipiens fatigans* Wiedemann (*quinquefasciatus* Say) replaces *pipiens* in the tropical and subtropical regions of the world, and in many areas its range overlaps that of *pipiens*." "No consistent biological differences between *C. p. pipiens* and *C. p. fatigans* have been found. The only morphological distinction between *pipiens* and *fatigans* is seen in the DV/D ratio, which is a measure of the angle of inclination of the ventral arms of the mesosome. In a recent study, Barr has shown that *pipiens* exists in North America as a cline; the DV/D ratio is smallest in more northern populations and increases progressively in the populations toward the south. In a broad belt from 36° to 39° N. lat., where overlapping of the two populations occurs, Barr found evidence of hybridization and mixing."

Another presumed species in this complex is *Culex molestus* Forskal, which "occurs in many localities within the *pipiens* range." "The only clear-cut difference between *molestus* and the other *pipiens*-group populations is in the ability of the females to lay eggs without having fed in the adult stage. Such autogenous eggs are produced from food reserves accumulated in the larval stage, and it has been suggested that the richness of the larval diet would be the determining factor." "The problem of *molestus* populations and their place in the taxonomic picture is an extremely complicated one. *Molestus* is morphologically similar to *pipiens*, but very dissimilar physiologically; both *pipiens* and *fatigans* seem to be further removed physiologically from *molestus* than they are from one another. Some populations of *molestus* will interbreed with *pipiens* or *fatigans* or both, but widespread sterility exists among different populations of *molestus*. One might argue that the *molestus* phenotype actually represents a fairly common recombination of genes already present in *pipiens* populations. This combination of genes might then be selected by a specific type of environmental situation, producing *molestus*. This argues of course for the widespread distribution of '*molestus*' genes throughout the *pipiens* populations. In modern genetic terminology, '*molestus*' would represent isolated populations in which the frequencies of certain genes were much different than in the normal *pipiens* population. Continued isolation and presumably selective advantage of the 'biotype' might eventually permit the accumulation of cytoplasmic differences which in turn take over as isolating mechanisms. If this be the case we are dealing with an unusual and interesting type of mechanism of speciation."

The results of taxonomic work on aphids and gall wasps indicate that detailed investigations, which would have to consider extensive geographic areas, would probably show that similarly complicated situations are not rare exceptions in other groups of animals. The usual concepts for infraspecific categories scarcely seem adequate for presenting such complex situations, and we have therefore not discussed them fully in all their variants. In such cases there is no alternative but to discuss the complex structure of the particular species in all possible detail, instead of forcing the presentation into a fixed system of categories as was particularly worked out by Paramanow (1944).

THE TAXONOMIC TASK IN THE AREA OF THE HIGHER GROUP CATEGORIES

The Mode of Origin of Higher Taxa and the Question of Their Real Existence

In the preceding chapters it was shown that all species that have ever lived, when arranged according to their phylogenetic relationships, correspond to the definition of a hierarchy (partitioning hierarchy) in the sense of Woodger and Gregg. Accordingly the adequate form of presentation is the hierarchic system. The questions that we must now pursue in detail result from the structure of such a hierarchic system.

Two different graphic representations of a hierarchic system are contrasted in Fig. 18. Comparison shows that they correspond exactly to one another: every fact that can be inferred from I is also expressed in II; the only difference is the way in which the phylogenetic relationships of the individual species are expressed. In II the phylogenetic relationships between a species (stem species) and its successor species are indicated by arrows. In I they can be recognized equally well from the fact that the symbolic boundary lines of the stem species have been drawn around their successor species. The form in representation I has a particular significance because it shows how the boundaries of the higher taxa must be drawn in the phylogenetic system: in I the symbols for the "stem species" 1, 2, 3, 4 coincide with the boundaries of the higher taxa into which the species that arose from them are collected in groups in the phylogenetic system. From this it is evident that to every higher taxon in the phylogenetic system there corresponds a "stem species" from which all the species included in the taxon have arisen. It is also evident that in the phylogenetic system the species included in each higher taxon must be derivable from a common stem species, and that no species having arisen from this stem species can be placed outside this taxon.

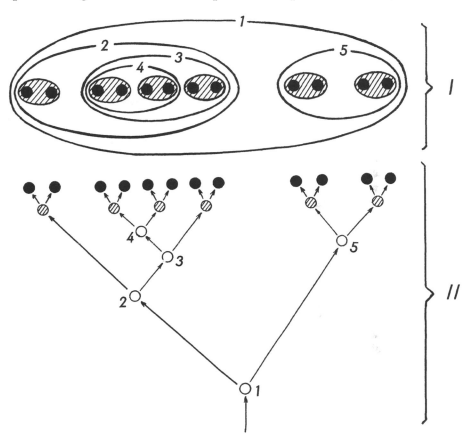

Figure 18. The phylogenetic kinship relations between the species of a monophyletic group, represented in two different ways.

From the fact that in diagram I the boundaries of a "stem species" coincide with the boundaries of the taxon that includes all its successor species, it follows that the "stem species" itself belongs in this taxon. But since, so to speak, it is identical with all the species that have arisen from it, the "stem species" occupies a special position in this taxon. If, for example, we knew with certainty the stem species of the birds (and it is only from such a premise that we can start in theoretical considerations), then we would no doubt have to include it in the group "Aves." But it could not be placed in any of the subgroups of the Aves. Rather we would have to express unmistakably the fact that in the phylogenetic system it is equivalent to the totality of all species of the group. Naturally, in practice this meets basically insurmountable difficulties because it is scarcely ever possible to determine with certainty whether one (and in this case which) of the known species of *Archaeopteryx* (to continue with our example) is the stem species of all the other known species of Aves.

Such difficulties do not arise in a system that represents only the species of a particular time horizon, because here there are no stem species. The problem of the so-called "living stem species" in neozoology is an artefact of a morphological or partly morphological species concept. As our knowledge of fossil animals increases, even paleontology is more and more forced to restrict systematic monographs to the species of a particular time horizon. Then the problem of "stem species" does not arise.

It must be emphasized that these are practical and not theoretical difficulties that face the arrangement of the species of different time horizons in one and the same system. They do not even concern all fossil species, but in principle only those of which it cannot be determined whether or not they are stem species of species groups of later time horizons. These difficulties can be largely avoided if the systems of different time horizons are kept apart—a procedure to which, as we said, paleontology is more and more being forced.

Comparison of the two diagrams in Fig. 18 shows that time of origin must be regarded as the measuring stick for (relative) ranking, the subordination of the taxa in the hierarchy. This necessarily follows from the fact that the phylogenetic system is a "partitional hierarchy," with the species as the unit that divides. Bigelow (1956) correctly states that the rank order (subordination) of the higher taxa in the phylogenetic system rests on "recency of common ancestry."

It is often said that only monophyletic groups have any place in the phylogenetic system—in other words, that the phylogenetic system should be a portrayal of all determinable monophyletic groups in the animal kingdom. This definition is valid, as an expression of the facts we have formulated in other words, only for a specific definition of the concept "monophyly" and "monophyletic group."

The definition of Mayr (1942) probably represents a widely held view: "We employ the term monopyletic as meaning descendants of a single interbreeding group of populations, in other words, descendants of a single species." But this definition is incomplete; it would admit, for example, a group in which birds and mammals are included under the name Homotherma. Such a grouping is in fact defended by Huxley (1958), although he obviously knows and emphasizes that "homothermy" was achieved independently by birds and mammals. No doubt at some time in the Paleozoic there existed a species from which all mammals and all birds have descended. Huxley's Homotherma would thus be a mono-phyletic group in Mayr's sense, but Mayr himself probably did not intend to make such a conclusion possible. Bigelow (1956) correctly says that "unless the time element is introduced, say as follows: 'the members of a monophyletic group share a more recent common ancestry with one another than with any member of any other such group of equal categorial rank,' the term 'monophyletic' is meaningless."

But even Bigelow's definition is not free from objections, because it does not stipulate that "common ancestry" must mean a common stem species (in the sense of the genetic definition of species); also the concept "group of equal cate-gorical rank," although perhaps not falsely meant, is misunderstandable and at any rate superfluous in this connection.

From what was said above concerning the erection of higher taxa in the hier-archic system of phylogenetic systematics (without using the term "monophyletic group"), it is evident that these higher taxa can be called "monophyletic groups" only if this concept is defined as follows: A monophyletic group is a group of species descended from a single ("stem") species, and which includes all species descended from this stem species. Briefly, a monophyletic group comprises all descendants of a group of individuals that at their time belonged to a (potential) reproductive community, i.e., to a single species. This definition makes it possible to label as monophyletic even groups of animals that do not have bisexual repro-duction if it can be shown to be probable that these groups descended from stem species with bisexual reproduction.

In all these definitions it is important that we can speak of a "monophyletic group" if it can be shown not only that all species (or individuals) included in it actually descended from a single stem species, but also that no species derived from this stem species are allocated outside the group in question. Naturally this does not mean that the birds, for example, cannot be called a monophyletic group if the fossil species are disregarded in a presentation of the system. They are not classified outside the group "Aves," but are simply disregarded for a spe-cial reason.

Another definition is: A monophyletic group is a group of species in which every species is more closely related to every other species than to any species that is classified outside this group. This definition contains the idea of "kinship," which is also present if we say that in the phylogenetic system the species are classified according to the degree of their kinship. The correctness of these defini-tions depends on the definition of the concept "kinship." Unfortunately the use of this concept in systematics is a source of endless confusion and controversy that is obstinate and superfluous.

Cain and Harrison (1958) represent the viewpoint that "affinity in practice means degree of overall resemblance estimated from available characters." In their opinion, "affinity in a phylogenetic sense is both logically and historically posterior to that of affinity as overall similarity." At least it is scarcely true that the German concept of *Verwandtschaft* historically has meant first "overall similarity." Rather it was originally a genetic concept. Only later was it transferred from the realm of the tokogenetic relationships of individuals to that of phylogenetic relationships on the one hand, and to "similarity" on the other. Actually, similarity is usually particularly evident in tokogenetic and phylogenetic "relatives."

The asserted logical primacy of the morphological concept of kinship is not true either. It rests on the erroneous assumption that "phylogenetic relationship" also is to be recognized only by determining "affinity" or "kinship" in the sense of "overall resemblance." We have already rejected this assumption (p. 12), and will treat it in more detail in the sections on morphological methods in phylogenetic systematics.

The concept of "relationship" may be defined as follows: "A species x is more closely related to another species y than it is to a third species z if, and only if, it has at least one stem species in common with species y that is not also a stem species of z." This definition agrees with that of Zimmermann (1931): "The relative age relationships of the ancestors is the only direct measure of phylogenetic relationship." Using Bigelow's definition, we can also say that the measure of phylogenetic relationship is the "relative recency of common ancestry."

The definition of (phylogenetic) relationship can be broadened to make it applicable to monophyletic groups of any rank: A particular taxon B is more closely related to another taxon C than to a third taxon A if, and only if, it has at least one stem species in common with C that is not also a stem species of A. When so defined, the species and monophyletic groups of higher rank in the phylogenetic system are ranked according to the degree of their kinship. But what happens if the concept of kinship or "affinity" is equated with "overall resemblance" in the sense of Cain and Harrison (1958)? Can we then say that species are grouped in the phylogenetic system according to the degree of their kinship (affinity), and that a group in which any species is more closely related to any other species than to any species outside this group is a monophyletic group?

In discussing the morphological methods of phylogenetic systematics we will see that by making certain assumptions it is perhaps possible to measure the "overall resemblance" of species, and even to arrange the species in a hierarchic system on the basis of these measures. This is possible, however, only with the aid of a device that will be discussed in detail later on. Furthermore, it has not been proved, and is even improbable, that the arrangement of species in a hierarchic system with the aid of this device actually expresses the overall resemblances of the species in the best possible way. At any rate it is also possible to present relationships of morphological similarity in a hierarchic system, although this hierarchic system is not a partition hierarchy like that of phylogenetic systematics. In our diagram of the phylogenetic system (Fig. 4), the symbols for the genetic species would represent morphological correlation coefficients, and the arrows would merely connect correlation values of different magnitudes. They

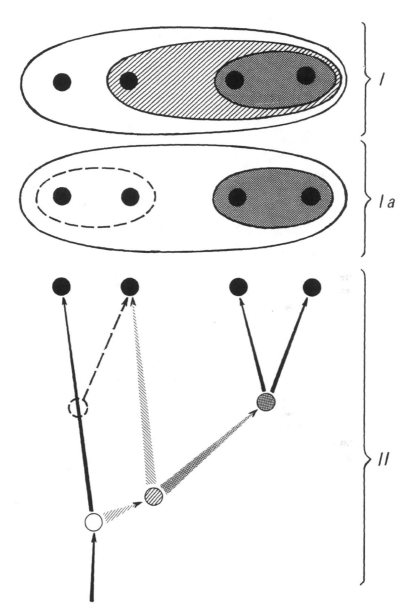

Figure 19. Diagram to show that the hierarchic type of system is suitable to represent phylogenetic relationships only if this concept is defined as specified on p. 74.

would not indicate real cleavage processes, but only a particular method according to which the correlation values have been ordered.

It is generally agreed today that there is no firm relationship between the degree of morphological similarity (overall resemblance, static relationship, form relationship) of species and the degree of their phylogenetic relationship (as defined above). None of the authors who have occupied themselves with measuring morphological similarity have maintained that there is. Consequently the concepts of "phylogenetic kinship" and similarity (form relationship) must be kept strictly apart. Unfortunately, even today this is not always done. "In discussions of dendrograms and their construction, confusion between phyletic and static relationships pervades much of the literature" (Michener 1957). "It is not always easy to determine, for example, whether a given author means 'similarity' or 'recency of common ancestry' when he uses the term 'affinity' " (Bigelow 1958).

This equivocation of terms is connected with the fact that the hierarchic type of system is used for representing both phylogenetic kinship and the entirely different form relationship. Such ambiguity is dangerous because it greatly favors the logical error of metabasis in conclusions drawn from the structure of the classification of an animal group. Fig. 19 shows what this means. A taxonomist who works according to the principles of phylogenetic systematics will always be inclined to interpret the hierarchic system of any author who says he intended to present the "affinities" of the species as a partition hierarchy in the sense of phylogenetic systematics. Consequently he will assume that the species A and B (Fig. 19, Ia), which the author of the system united in a group, form a monophyletic group. But actually the author of the system only intended, by including the species A and B in one group, to express their form relationship (affinity in the sense of overall resemblance). Consequently only a certain value of the correlation coefficients of their characters connects the two species A and B in his hierarchic system, rather than the possession of a common stem species. Serious fallacies may result from the use of such a grouping, for example in zoogeographic investigations. Furthermore, in controversies between two authors as to the "affinity" (*Verwandtschaft*) of species B, there can never be agreement if to one author "affinity" means phylogenetic relationship, and to the other it means form relationship (overall resemblance).

The situation becomes completely bewildering if an author explicitly recommends using both concepts of relationship in constructing one and the same system: "In spite of their objectivity, it has seemed best not to use static relationships alone to construct a formal classification because it would be quite different from currently accepted classifications of many groups and because phylogeny has been the usually accepted basis for classifications in zoology. It has seemed preferable to me in the past to use presumed phyletic relationships as the main basis of classification, considering as secondary factors static relationships, number of species per group, and gap size" (Michener 1957). We have already explicitly recognized (p. 4) that in systematics there can be different systems corresponding to the different kinds of relationships that exist between organisms. To this extent we can agree with Michener's statement that "static and phyletic relationships are of equal interest to systematic biologists." But combining different systems in a syncretistic system robs the combination of any scientific value, since it could never be used as the basis for investigations that presuppose knowledge of a particular kind of relationship between the organisms. "Classification must be based on one or the other [on "overall resemblance" or "recency of common ancestry," on "form relationship" or "phylogenetic kinship"], not on both, if philosophical confusion is to be avoided" (Bigelow 1956). We would add that not only philosophical confusion, but also false conclusions (by way of the logical error of metabasis) about the course of evolution are the consequences of mixing the principles of classification—as is, for example, explicity recommended by Stammer (1961). Such false conclusions are evident everywhere in the literature.

A general question that remains to be answered before we turn to the special methods of phylogenetic systematics is the reality of the different group categories of the system. Opinions are still divided on this. As representatives of two strongly opposing views that are still defended, Möbius and Nägli may be quoted (both quotations from Möbius 1886).

Möbius contends that for those who assume that all organisms developed from a single primary form "only the individuals are real, and the species and all higher systematic groups are only concepts that express relationships that are really represented by the simultaneously or successively existing individuals." He says further: "all complexes are thought-connections at the basis of which lie concrete things—the visible individuals. The doctrine of descent, guided by ideas, goes beyond experience." According to Plate (1914) a similar viewpoint is represented by Agassiz, Claus, Bessey, Jest, and more recently by Martin (1929) and Diels (1921), who says that all attempts to prove that any categories are real have failed. Nägli, who is explicitly opposed by Möbius, says that "the center of gravity in natural history no longer lies in the species, but in the fact that each systematic category is taken to be a natural unit representing the point of penetration of a great evolutionary movement. The genera and higher concepts are not abstractions, but concrete things, complexes of forms that belong together, that have a common origin." According to Plate this viewpoint is also represented by Heincke, and more recently by Beurlen (1936) in a more or less clear-cut fashion. Interpretations that lie between these two extremes are also frequent. Haeckel (1866) is probably relatively isolated:

As the only real categories of the zoological and botanical systems we can accept only the great major subdivisions of the plant and animal kingdoms, which we have called stems or phyla and have discussed as genealogical individuals of the third order. In our opinion each of these phyla is a real unit of many forms that belong together, since it is the material bond of blood relationship that embraces and unites all members of each phylum. All the different species, genera, families, orders, and classes that belong to such a phylum are continuously interconnected members of this larger and more inclusive unit, and have gradually developed from a single primary form.

Most of the older and more recent authors are of the opinion that both the individuals and the species are real, whereas the higher taxonomic categories are mere abstractions. For Plate (1914), for example, the species occupies "a position distinct from the genus, family, etc. in that it exists in nature as an actual 'complex of individuals' independent of human analysis, and therefore as an objective entity. The members of a species recognize each other and reproduce together, whereas the higher groups of individuals (genus, family, etc.) are not formed through themselves, but by the comparing and reflecting mind of man. In this sense the species is real, whereas the genus, family, and other higher groups are abstractions." (quoted from Uhlmann 1923). According to Plate, the same view is held by Delpino, Kerner von Marilaun, Balli, Burmeister, Dana, and Brauer. The same view is defended by Rensch (e.g., 1934), who distinguishes the species, rassenkreise, and lower categories as "objective" from the "subjective" higher categories. This is likewise true of Kinsey (1936, 1937), and the Claus-Grobben-Kühn textbook of zoology, widely distributed in Germany: "no real object corresponds to the higher systematic categories, from the genus up."

In view of these great differences of opinion, which are still unresolved, it is understandable that Simpson (1951) maintains that such concepts as "real," "natural," and "objective" should not be used for taxonomic categories and methods of classification. But such resignation is scarcely advisable, because the systematist obviously would like to know whether the methods he is using give objective or subjective results, and whether the categories of his system have "real existence" or not, whether they have the character of individuality or are generalizations. Furthermore, if disputes about whether a phylogenetic or a topological system is possible at all, whether one is logically dependent on the other, and whether one or the other should be made the general system of reference in biology, are carried on with arguments in which such concepts play a decisive role, then how can the systematist decide in favor of one or the other if he declines to familiarize himself with the concepts? He has to decide, if he is to avoid the justifiable charge of not knowing what he is doing or even what he is trying to do.

These differences of opinion seem to me to have their main source in the erroneous identification of logical systematics with biological systematics. This is especially evident in the statements of Möbius and Plate quoted above. Collier (1924) also believes that the goal of systematics is "to construct the logical arrangement of the multiplicity of forms." According to Diels (1921), "as in any logical classification, the objects that are put together in a group must have more characters in common than they have with any other objects excluded from the group." According to Zündorf (1940), for Steiner (1936) too "systematics is without any doubt a logical relation."

Martini (1929) also expressed this viewpoint very clearly. Having recognized that only individuals, and not species, exist in nature, he asks: "What then are the species? They are concepts (generic concepts in the philosophical sense) that we ourselves produce. . . ." Thus the group categories of the biological system are identified with the general concepts (generic concepts), the universals of logic. This would lead the controversy over the reality of the supra-individual groupings of the biological system back to the controversy over universals that has played an important part in history, especially in the philosophy of the Middle Ages (existence of universals *ante rem, in re,* or *post rem*). This identification is favored by the use of some of the terms of logical systematics for group categories in biological systematics.

Linnaeus "took from the old logica materialis the concept of genus, species, and difference" (Thompson 1952). Especially well known is the definition of the concept "definition," in which the conformity is clear: "*Definitio fit ex genere proximo et differentia specifica.*" Thompson therefore concludes that "since the species is not the individual or the collectivity of individuals, it does not exist, as such, in the real world. It can therefore exist as such only in the mind." "The object of the species definition," according to Thompson, is "an abstraction which has nevertheless a foundation in the individual natures. It is not therefore a figment of the imagination, neither is it a pure idea having no relation to the material world. It is an abstraction having nevertheless a foundation or source in nature since it is realized in the individuals representing the species." According to Thompson the categories genus, family, etc. are "merely what the medievals called 'metaphysical degrees.' Like the species, they are abstractions, each category having a greater extension, as the logicians say, than those below it, while on the other hand it is less extensive." Thus according to Thompson the categories above the individual in the biological system are general concepts developed only by abstraction; they have only intentional, not real, existence. "With the concepts of intentional and real existence we also designate the difference in nature that separates the general concept from the concrete individual." "Only the individuals have real existence. . . . The general concept has a '*fundamentum in re.*' Its content is realized in the individuals, and only in the individuals."

According to Thompson the categories of the biological system also have in common with the general concepts of logic the fact that they are the poorer in characters the higher the category rank. "Thus the category Hymenoptera is more extensive than the categories Ichneumonidae and Braconidae, since it includes both, insofar as they have something in common; but it is less comprehensive, since it excludes what is special to each group. . . . For example, it has wings of no definite venation, an ovipositor of no definite form, amorphous mandibles. . . ."

Thompson's error in equating logical systematics with biological systematics is clear from this example. This equivalence may be true for morphological or typological systematics, but in no case is it true for phylogenetic systematics. The categories of phylogenetic systematics are not constructed by abstraction. They are not defined as bearers of a complex of characters that remains if, starting with the individuals, we subtract more and more characters that are specific to the individuals and then to progressively more inclusive groups of individuals. In the phylogenetic system the categories at all levels are determined by genetic relations that exist among their subcategories. Knowledge of these relations is a prerequisite for constructing the categories, but the relations exist whether they are recognized or not. Consequently here the morphological characters have a completely different significance than in the logical and morphological systems. They are not themselves ingredients of the definition of the higher categories, but aids used to apprehend the genetic criteria that lie behind them.

Perhaps a simple example will best show that there is a fundamental difference between phylogenetic systematics and logical or morphological systematics. There are still many families that were torn apart by the war. The members of these families know nothing of each other. Children have grown up in the meantime. No one doubts that such families exist. Their boundaries as "supra-individual groups" are determined by the tokogenetic relationships (Fig. 6) that exist between the members of the families. Now, there are organizations that try to reassemble such families. This can be done only by discovering the tokogenetic relationships. Even if one is dependent on indirect methods, and perhaps has to check and decide the family affiliation of certain children—who have grown up in the meantime—by hereditary analysis of "holomorphous" characters, the reassembling of the family is not constructing a group by logical abstraction of characters as in the construction of general concepts and in purely morphological systematics.

The work of phylogenetic systematics consists of such "family reassembly." This is not altered by the fact that the structure of the phylogenetic relationships is different from that of the tokogenetic relationships (Fig. 6). It is noteworthy that even representatives of modern logic (symbolic logic, logistics) have correctly recognized this difference. Unfortunately neither Thompson nor Blackwelder (1959) has taken note of this. Woodger (according to the definition of Gregg 1954) has developed a simple language "with a structure entirely different from that of set theory, in which taxonomic group names may be construed as names of individuals." Woodger (1952) proceeds from the example of a square, which can be subdivided into smaller squares. "If 'X' names each of these smaller squares, then ΣX names the larger square of which they are parts." Consequently this is a hierarchy (division hierarchy), as in the system of phylogenetic systematics. In such a hierarchy, according to Woodger, the higher categories are not "sets of organisms," but the subordinate categories are "parts" (in the true sense of the word) of the higher ones. According to Woodger the "evolutionary species and genera" are not abstractions like the categories of logical and morphological (unfortunately he calls them "taxonomic") systematics, but "concrete entities with a beginning in time." He says there is in this respect no difference between the species and the higher categories (in the phylogenetic sense). He rejects the view that one can distinguish "species as real from genera as unreal." However, he does not commit himself on the concepts "real" and "individual" for characterizing the species and higher categories (as Gregg does in his remarks on Woodger's works). He simply calls them "concrete entities with a beginning in time," in contrast to the "abstract, timeless" categories of morphological systematics.

But according to N. Hartmann (who is followed almost without reservation by Max Hartmann in his work on the philosophy of the natural sciences), temporality is the only characteristic of reality and individuality: "The true characteristics of reality are not dependent on the categories of space and matter, but on those of time and individuality. And temporality is inseparably connected with individuality. It consists in nothing else but the onceness and the singleness" (N. Hartmann 1942). By reality Hartmann understands "the mode of existence of everything that has a place or a duration in time, its origins and its cessation."

That in any case the space-matter (bodily) relationship of the parts is not decisive for the concept of individuality and reality is also emphasized by other authors. Ziehen (*loc. cit.*, I) illustrates this with reference to the small planets, von Bertalanffy with reference to a swarm of bees: "We call the bee an individual organism because it represents an order of magnitude of things that appears to us 'with the unaided eye' as a distinct individual thing. A swarm of bees may likewise appear to be an 'individual thing' at a distance, and we perceive that it is composed of a great number of individual things only by closer examination."

Finally, even substantial constancy is not a prerequisite for recognizing an individual as "the same" at different times. "We ascribe substantial sameness to two successive R's, R_1 and R_2, if the difference between R_1 and R_2 is causally understandable through a continuous sequence of intermediate states" (Ziehen, I). Only on the basis of this criterion can we recognize, for example, the acorn and the oak tree that develops from it as one and the same individual: "thus two phases of the same thing may be completely different in all their R-components [in this sense, peculiarities or characters], and yet on the basis of the continuous causal connection we speak of a single thing" (Ziehen, I).

If we now attempt to evaluate the categories of the phylogenetic system from the viewpoint thus gained, there can be no doubt that all the supra-individual categories, from the species to the highest category rank, have individuality and reality. They are all (Fig. 13) segments of the temporal stream of successive "interbreeding populations." As such they have a beginning and an end in time (N. Hartmann), and there is a constant causal connection between the phases in which they are found at different times (Ziehen). All this is missing in the categories of the morphological or typological system, which consequently are timeless abstractions (Woodger) and therefore have neither individuality nor reality.

On the other hand, it cannot be overlooked that there are essential differences between the individuals, which exist as such even in everyday language, and the "supra-individual," particularly the supraspecific, taxonomic categories. These differences are probably not merely of a gradual nature in the sense that, measured with the yardstick of our human proportion relations, one is given immediately as individuals whereas the other must first be determined as such. On the contrary, it is often emphasized (N. Hartmann 1942, for example) that we cannot assume that the laws ascertained for a particular level in the graded structure of things are also valid for other (e.g., higher) levels. This is true even when things of the higher levels appear to be "composed" of components that resemble or are identical with the individual things of the lower levels. Now there is no doubt that the higher taxonomic groups, like the species, consist of individuals in the strict sense of this word, and that they therefore represent a higher level of more complex things than the species do. Consequently, if we transfer the concept of individuality prematurely to the higher taxa, we expose ourselves to the criticism of making it impossible—or at least difficult—to recognize differences in the laws that are valid. If we look for differences between individuals (in the strict sense) and the higher taxa, then in addition to the difference in the coherence of the components the following also applies (von Bertalanffy, I): "We can speak of individuality in only one sense, namely that ontogenetically and phylogenetically an increasing unification takes place." "Thus individual denotes unification of components through organic relations into higher operational units, with these relations becoming increasingly firm as we ascend in the ontogenetic and phylogenetic sequence, the individual parts becoming increasingly differentiated and less independent."

Thus the most important thing in von Bertalanffy's statements concerning the concept of individuality is seen to be that the individuals represent "operational units." Probably we must interpret this concept to mean that an individual is a unit both externally (in its effect on others), and internally (in the reciprocal action of its components). In this respect there is unquestionably a distinct difference between what in systematics are simply called individuals and the "supraindividual" categories, to which N. Hartmann and others also attribute individuality. This difference is most evident if we begin with a unicellular individual, which as an "operational unit" in von Bertalanffy's sense shows all the characters of the individuality concept. Any protozoan is such a unicellular individual. If a protozoan divides, the products of the division and all their descendents resulting from further divisions are also individuals with all the characters of this concept. All of them together (a "clone") satisfy the definition of a divisional hierarchy. In their totality as a clone they also possess individuality and real existence in N. Hartmann's sense, but the clone lacks the character "operational unit." On the other hand, the zygote of the Metazoa is also a cellular individual with all the characters of individuality, including that of operational unit. It too divides, and the division products and their descendents are again cellular individuals, which at least in many cases retain all the characters of the individuality concept. In their totality they too form a divisional hierarchy. But in contrast to the individuals of a clone they remain attached to one another, and together form an organism that itself is an operational unit in addition to having the other characters of individuality. In this example the difference between an individual in the true sense and the categories of a divisional hierarchy is clear: the latter lack the character "operational unit."

It is more difficult to determine the particular individuality character of the species. There can be no doubt that, like the higher categories of the phylogenetic system or of any other divisional hierarchy, they have "place or duration in time." It is questionable, however, whether the species can be regarded as operational units within their environment. That there are forces acting "inwardly" to hold their components together follows from the definition of the species as a "reproductive community" that has a pool of genes harmoniously adjusted to each other. At least this is true in those cases in which a species is a true closed population and reproductive community. But those categories—also called species—that consist of complexes of incompletely isolated vicarying reproductive communities form a transition to the categories of higher rank.

Thus all categories of the phylogenetic system are characterized by individuality and reality, in contrast to the abstract and timeless categories of the morphological system. This does not mean that they are all "individuals" in exactly the same sense as the individual organisms that are the units of life, and which we ordinarily call individuals. Depending on the purpose we have in mind, we may emphasize the common traits in the mode of existence of the individuals (in the customary, strict sense) and the supra-individual systematic groups, or we may emphasize the differences. But we cannot transfer uncritically all the criteria of individuality that characterize one category rank to other category ranks if we are to avoid faulty conclusions.

Our conclusion that all categories of the phylogenetic system have individuality and reality is true, of course, only if our system reflects accurately the divisional hierarchy to which its elements belong in nature—in other words, if the genetic relations between them are correctly recognized. Consequently we must now turn to the question of whether, to what extent, and with what degree of exactness the methods available to us permit fulfilling the theoretically posed task of phylogenetic systematics.

Taxonomic Methods in the Higher Group Categories

The hierarchic system used in phylogenetic systematics is composed of monophyletic groups (Figs. 2, 6). These groups are subordinated to one another according to the temporal distance between their origins and the present; the sequence of subordination corresponds to the "recency of common ancestry" of the species making up each of the monophyletic groups. But this expresses only the relative ranking of the categories. In addition there is the question of the absolute rank of the individual categories. The rank sequence class—order—family—genus merely denotes the sequence within one and the same category. It tells us nothing about the criteria by which a particular category must be designated a class, rather than an order or a family.

The result is a dichotomy in our task. We have to determine what methods are available for establishing the relative ranking of the group categories, and then we must consider the principles according to which the particular names are assigned to the individual categories to give them a particular absolute rank as class, order, etc.

THE DELIMITATION OF HIGHER GROUP CATEGORIES AND THE DETERMINATION OF THEIR RELATIVE RANK ORDER

A number of more or less reliable and well worked-out methods available for determining the phylogenetic relationships between the species and the higher taxa will be discussed below. The comparative holomorphological, the paleontological, and the chorological methods (or groups of methods) may be distinguished.

The Comparative Holomorphological Method. Morphology, or more correctly comparative holomorphology, occupies a central position in phylogenetic taxonomy because critical investigations of the possibilities and capabilities of phylogenetics have been concerned to an overwhelming degree with morphological methods. Often discussions of such investigations are accompanied by an explicit assertion that the determination of phylogenetic relationships is restricted to their derivation from relationships of morphological similarity. Even discerning and distinguished modern authors seem to assume that the morphological methods of phylogenetic systematics are limited to inferring the degree of phylogenetic kinship from the degree of morphological similarity between species.

"On the average, two animals with more homologous characters in common are more nearly related, their ancestral continuity is relatively more recent, than two animals with fewer. . . . The rule that degree of homology is directly proportional to degree of affinity is true within limits narrow enough for most purposes and is a valid working principle" (Simpson, cited from Bigelow 1958). "The categories of the systematist are based on the degree of similarity, and the more closely two organisms are related phylogenetically, the more morphological characters they will usually have in common" (Mayr, in Hedberg 1958).

Thus Bigelow's case against phylogenetic systematics is partly based on showing that "similarity will not correspond with recency of common ancestry in the majority of cases." (See also Bigelow 1959.)

The Absolute Degree of Similarity and Its Relation to the Degree of Phylogenetic Relationship. The simplest form for inferring degree of phylogenetic relationship from the similarity relations between species and species groups is the equation: "the more similar, the closer the phylogenetic relationship." We have already tried to show (p. 14) that the possibility of deriving the theory of descent from systematics can be understood only because in most cases this equivalence is actually true. But the fact that there is no yardstick for measuring degree of morphological similarity, and that already in the oldest systems the use of genetic criteria resulted in a partial even though unconscious evaluation according to phylogenetic principles, urges caution with regard to this interpretation.

A closer examination of the equation "similarity = common descent" finds its first difficulty in the fact that there is no commonly accepted way of judging similarity in form. Repeated attempts have been made to put the quantitative analysis of similarity relations on an exact basis. Naturally the methods of variation statistics and correlation analysis were primarily used. Smirnov (1923-38) saw his task as a "geometrization of organic form." As a "correlation equation of race" he uses the Edgeworth-Pearson function, in which the correlation of characters is expressed. "By applying the Edgeworth-Pearson function not only to races, but also to congregations of any order, we solve the basic problem of systematics: the exact characterization of systematic categories." Later Zarapkin (1939) tried to measure the "divergence between the races and species." More recently McGuire and Wirth (1958), Sokal and Michener (see Sokal and Michener 1952, Michener 1957, Sokal 1958, and Ehrlich 1958), and Cain and Harrison (1958) have developed and recommended mathematical methods for measuring morphological differences between species.

It is regrettable that even the most recent authors present and recommend their own methods but never explain why their method deserves preference over those of their predecessors. Consequently it is almost impossible for a systematist with inadequate mathematical training to judge whether any progress has been made in measuring differences in form. According to Bavink (1941), a prerequisite for such progress is "to shift the basic concept of all existing mathematics . . . to place the concept of measurable and countable 'quantity' in second place, and the basic biological concept of 'form or gestalt' in first place, and recognize that the former is a limiting case of the latter." "If this plan were actually carried out, it would obviously require a new mathematics." Bavink finds the germs of such a "mathematics of form" in the theory of integral equations, combinatorics, and calculation of variations, but these germs relate to what is attempted in the same way that the problem of tangents, calculations of content, etc. do to true infinitesimal calculation.

Accordingly it is understandable that the calculation of variation and correlation is particularly successful in the characterization of closely related groups (groups of individuals, races). These lie close to the boundary where, according to Bavink, the form and gestalt concept passes over into the measurable and countable quantity that is graspable with existing mathematical methods. Consequently in this area differences in form can be represented numerically with adequate approximation.

According to Bavink the development of a mathematics of form from the present beginnings would require a "mathematical genius of our day," who would have to accomplish "what Newton and Leibniz accomplished in their day." Even if one of the currently recommended methods for measuring differences in form represented such a stroke of genius, phylogenetic systematics would be interested only in the question of whether and how far the differences in form so recognized would determine degrees of phylogenetic relationship. The first prerequisite for this is a method of representing adequately the relationships of morphological similarity (overall similarities) of the species in a hierarchic system. This is necessary because a hierarchic system is the adequate representation of phylogenetic relationships. The question of whether the two systems are identical can be asked only if the relationships of morphological similarity can likewise be represented completely correctly in a hierarchic system.

Now, it has long been a recurrent contention that the relationships of morphological similarity correspond to the picture of a network rather than to a hierarchic system. Horn has repeatedly noted (e.g., 1929) that Linnaeus used

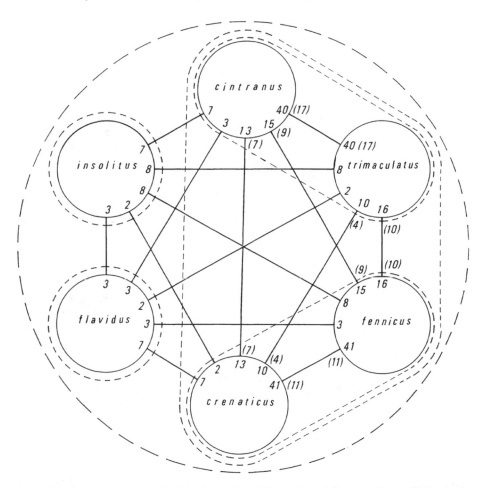

Figure 20. Diagram of the typological relationships of the species of the genus **Cyrnus** (Trichoptera). The numerals signify the characters common to the species. (After Klingstedt.)

the image of an interrupted network for this kind of similarity relationship between organisms, although it is to be emphasized that Horn's mixing of this
image with the idea of a polyphyletic origin of "species and genera" is completely
inadmissible. This reticular relationship is also emphasized in botany (Diels 1921,
Anderson 1937: "morphological relationships reticular rather than dendritic").
Helmcke's attempt to classify the brachiopods (Fig. 46) and the work of Klingstedt (Fig. 20) are examples.

Cain and Harrison (1958), who have developed a method of their own for
measuring form similarities (or differences) among species, acknowledge the
same view in clear words: "We can say that although the mutual differences
(and therefore similarities) of three forms can usually be represented by a diagram in two dimensions, those of four cannot necessarily be represented by the
diagram even in three dimensions, let alone two. Perhaps the use of additional
dimensions for every new form might help, but it seems doubtful." Nevertheless
these authors believe that the value of their method is that with its help it is
possible "to give a quantitative meaning to the taxonomic hierarchy." If the
M.C.D.'s (mean character differences) of a large number of species "are worked
out on a reasonably large number of characters, it would be possible to agree on
a M.C.D. value at the generic level, such that if a group is formed of species all
of whose M.C.D.'s are lower than this value, they are to be regarded as in the
same genus; but a species which has M.C.D.'s with the species in the group of
higher values must be put into a neighboring genus."

Obviously this method can be used without drawing arbitrary boundaries only
if there are recognizable gaps between the most similar groups. In other words,
it is applicable only if in multidimensional representation—which according to
Cain and Harrison is alone capable of symbolizing the mean character differences
of more than four species—there are "congregations" (aggregations of similar
forms). In these "congregations" (Smirnov) the mean character differences between any two species in each congregation must be less (i.e., the similarity must
be greater) than between any two or more species of any other congregation.
Two or more close congregations would then have to form groups of congregations of higher rank, for which the same would have to apply. Then there would
be an exact parallel between this hierarchy of communities of similarity and the
hierarchy of communities of descent (monophyletic groups) of the phylogenetic
system. But it would still remain to be proved that a community of similarity corresponds to a community of descent.

It has not yet been shown that the totality of all animal species, if arranged
according to the degree of mean character differences, actually gives a hierarchic
system of communities of similarity. It is not enough that such a system results
for a limited sector of the animal kingdom.

Cain and Harrison (1958) themselves state that "we would be able to choose
forms with any degree of difference" if all the species that lived in the geological

past were known. This probably means merely that there would then be no gaps between different communities of similarity.

If we assume that at least the species of a particular time horizon (the present, for example) can be arranged without forcing and without arbitrary boundaries in a hierarchy of communities of descent, then the assumption that the similarity groups of this system correspond to monophyletic groups of the phylogenetic system would be justified only if, in addition to a constant rate of evolution of all communities of descent, the extinction of species that produced the gaps between the communities of similarity had always taken place in such a way that the gaps within close communities of descent became no larger than the gaps that now separate them. This assumption is so improbable that it scarcely needs to be refuted, especially since paleontology has made no observations that support it.

The equation "community of similarity = community of descent" is also made impossible by the proven fact (see, for example, Simpson 1944) that there are differences in the rate of evolution (i.e., in the origin of morphological differences) in various communities of descent. "Unequal rates of evolution will produce cases in which overall basic similarity does not correspond with recency of common ancestry. Slow evolution in two phyletic lines will tend to produce an appearance of recent common ancestry when actual common ancestry is remote in time" (Bigelow 1958). Thus for this reason too communities of similarity will not always be communities of descent (monophyletic groups).

We may conclude that perfecting the methods of measuring overall basic similarity is of no significance for phylogenetic systematics. If absolute size of the differences in form is not a reliable measure of phylogenetic kinship, we must ask whether some other way of evaluating these differences is not a better means of discovering phylogenetic relationships.

The Rules for Evaluating Morphological Characters as Indicators of Degree of Phylogenetic Relationship. "Evolution is a transformation of organisms in form and mode of life through which the descendents become different from their ancestors" (Zimmermann 1953). Evolution in this sense (transformation) is also connected with speciation: if a species (reproductive community) is split into two mutually isolated communities of reproduction (Fig. 21, 1), there is always a change (transformation) of at least one character of the ancestral species in at least one of the two daughter species. In this simplest possibility, in the parent species and one daughter species there would be at least one character in the condition a, which would be present in a transformed ("derived") condition a′ in the other daughter species (Fig. 21, 1). This becomes of significance for phylogenetic systematics if the daughter species C, the bearer of the derived character a′, is split again into two successor species, D and E. Then there are two possibilities: a′ may be further transformed into a″ in one of the two new successor species (Fig. 21, 2, E) or a′ may remain unaltered in both successor species (D and C), while in one of them (Fig. 21, 3, E) another character b is

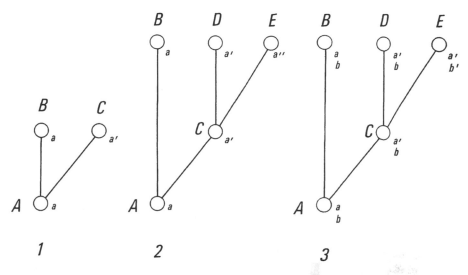

Figure 21. Speciation and character transformation.

transformed into condition b'. We assume that the character b was already present in the stem species A and remained unchanged in the first cleavage into species B and C. Consequently it will be present in condition b in the (recent) species B and D, but in the derived condition b' in species E.

In the following discussion it is often simpler if we call the character conditions a, a', and a" (or b, b', b") special characters. They are "characters" in the sense that they distinguish their bearers from one another, but we must always be aware of the fact that "characters" that can be compared are basically only character conditions that the real process of evolution produced by transformation of an original condition. Depending on the requirements of the problem, a, a', a", etc. will sometimes denote different characters, and sometimes different conditions of one and the same character, in the following.

We will call the characters or character conditions from which transformation started (a, b) in a monophyletic group *plesiomorphous*, and the derived conditions (a', a", b', b") *apomorphous*. Simple reflection shows that these are relative concepts: the characters a' and a" are both apomorphous compared with character a, but a' is plesiomorphous compared with a".

We will call the presence of plesiomorphous characters in different species *symplesiomorphy*, the presence of apomorphous characters *synapomorphy*, always with the assumption that the compared characters belong to one and the same transformation series. In our example (Fig. 21, 3) the species B and D are symplesiomorphous with respect to character b, the species D and E synapomorphous with respect to character a (Fig. 21, 3) or to the group of characters a′ and a″ (Fig. 21, 2). It is evident that the presence of corresponding characters in two or more species is a basis for assuming that these species form a monophyletic group only if the characters are apomorphous, if their correspondence rests on synapomorphy.

It makes no difference whether the synapomorphy consists in the fact that an apomorphous character (a′) is present identically in all species (Fig. 21, 3), or whether it is present in different derived conditions (a′ and a″). Recognition that species or species groups with common apomorphous characters form a monophyletic group rests on the assumption that these characters were taken over from a stem species that only they share in common, and which already possessed these characters prior to the first cleavage.

The possession of plesiomorphous characters (symplesiomorphy) does not justify the conclusion that the bearers of these characters form a monophyletic group. They may, of course, have taken over these characters from a common stem species (Fig. 21, 3, the species B and D have so received the plesiomorphous character b from the stem species A). But this stem species need not have been common only to them; other species may have descended from it that are bearers of apomorphous conditions of the same transformation series (Fig. 21, 3, E). That a common stem form is shared by a group of species (a condition for a "monophyletic group," p. 73) can be proved only by means of synapomorphous characters, not with symplesiomorphous characters.

The following discussion is based on the simplest case of three species or monophyletic groups; all more complex cases can be reduced to this. If the preceding considerations are used to determine the kinship relations of three species, it is entirely possible that one of them has only plesiomorphous characters (Fig. 21, 3, B). Synapomorphy must be shown for the other two species if the assumption that they form a monophyletic group is to be sustained.

On the other hand, if it is a question of determining the relationships between different species groups, then it is of primary importance to show that each group has apomorphous characters, characters that are present only in it. Otherwise it would be impossible to prove that it is a monophyletic group, and only as a monophyletic group can it be the bearer of phylogenetic relations to other monophyletic groups. The apomorphous features characteristic for a particular monophyletic group (present only in it) can be ignored in discussing its relations to other groups; we will call such characters the *autapomorphous characters* of a monophyletic group. Naturally these characters are "autapomorphous" only if the group in question is compared with other groups; so long as we are trying to show that the group itself is monophyletic, these same characters would be "synapomorphous characters" of the species making up the group.

All these considerations are summarized in a "scheme of argumentation of phylogenetic systematics" (Fig. 22). For simplicity, only one autapomorphous character is assumed for each monophyletic group (A, B, C, D, and the groups connected by rectangles, B + C + D and C + D). Naturally there may be more;

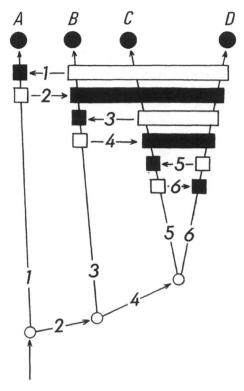

Figure 22. Scheme of argumentation of phylogenetic systematics. All groups regarded as monophyletic are distinguished by the possession of derived (apomorphous) stages of expression (black) of at least one pair of characters (synapomorphy of species of monophyletic groups).

the greater the number of autapomorphous characters that can be demonstrated, the greater the certainty that the group is monophyletic.

From Fig. 22 it is evident that in the phylogenetic system any monophyletic group, regardless of rank, must be characterized by the possession of autapomorphous characters. In the groups A, B, C, and D these are symbolized simply by black squares. In the groups B + C + D and C + D, black rectangles indicate that the autapomorphous characters (2 and 4) of these groups are at the same time the synapomorphous characters of their partial groups. Arrows connect the apomorphous characters with their plesiomorphous prior conditions, which are realized in other groups. Use of the same numbers indicates how the ties of phylogenetic relationship (1-6) are determined from the transformation series of the characters (1-6).

It is evident how this method of phylogenetic systematics, which is based on series of transformations of characters, differs from the previously discussed

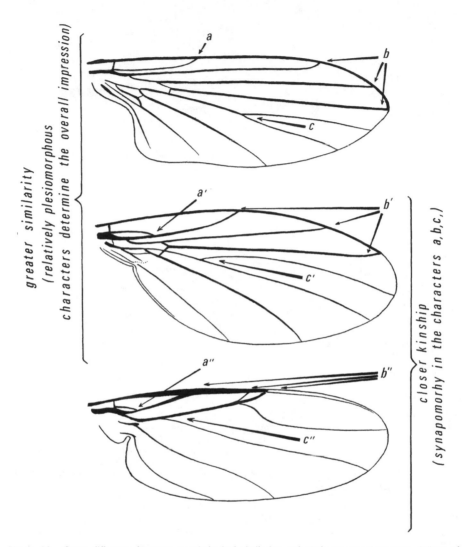

Figure 23. Great difference between morphological similarity and phylogenetic relationship. Wings of the dipteran families Clythiidae (above), Sciadoceridae (middle), and Phoridae (below). Synapomorphy of the Sciadoceridae and Phoridae: sc shortened and outlet into r_1 (a', a''); r_1 to r_4 thickened and outlets shifted basalward (b', b''); fork of the media moved near the radius (c', c'').

method of using absolute measurements of structural differences. When we distinguish different conditions of a character (a, a', a'') as plesiomorphous and relatively apomorphous, we do not take the magnitude of the differences between these conditions into account in any way. It is entirely possible for condition a' to differ far less from a than from a". The method that attempts to deduce the degree of phylogenetic relationship from the absolute magnitude of morphological similarity would in this case infer a closer relationship between a and a'. For our method, however, it is decisive that a' and a" are synapomorphous in comparison to a, and consequently we conclude that the bearer of a' is more closely related to the bearer of a" than to the bearer of a. This is shown in a concrete example in Fig. 23.

Many authors call the recognition that only synapomorphy, not symplesiomorphy, can be the basis for monophyletic groups "commonplace." Others, including those who firmly believe they are helping erect a phylogenetic system, explicitly recommend that species be grouped according to the criterion of symplesiomorphy: "The Cidaroidea, however, retain in stable condition many primitive features of Paleozoic aspect and origin, so that it seems more logical to group them with the Paleozoic forms in a subclass" (Durham and Melville 1957). Aczél (1951) unites several genera of the subfamily Taeniapterinae in a tribe of their own on the explicit basis that they possess a very plesiomorphous ("primitive") character.

The considerations that led to our "scheme of argumentation" of phylogenetic systematics disregarded the difficulties often encountered in practice in using the simple rule that "only synapomorphy justifies the presumption of monophyly in a group of species." These difficulties can be grouped around several questions. (1) How does one determine which characters of different species must be regarded as transformation conditions (a, a', a", etc.) of one and the same character (homology)? (2) How does one determine which is the beginning condition and which the terminal condition of a transformation series (a, a', a", etc.) (character phylogeny)? (3) Must the evolution of a character always progress in the same direction, or can a character revert to its original condition (a, a' a", a', a) (reversibility)? (4) Can certain transformation conditions of one character be attained by transformation of another character (a, a', a" and b, b', a') (convergence)? (5) Must a particular derived condition always be attained through a one-time series of transformations, or can the same condition be attained repeatedly from the same condition of origin $\left(\begin{array}{c} a, a', a'' \\ a, a', a'' \end{array} \right)$ (parallelism)?

We must now attempt to answer these questions.

(1) *Homology.*—Different characters that are to be regarded as transformation stages of the same original character are generally called homologous. "Transformation" naturally refers to real historical processes of evolution, and not to the possibility of formally deriving characters from one another in the sense of idealistic morphology. Since we can never directly observe the phylogenetic transformation of a character, the question arises as to what accessory criteria are available to convince us that particular characters are homologous in different species.

Apparently it is often forgotten that the impossibility of determining directly the essential criterion of homologous characters—their phylogenetic derivation from one and the same previous condition—is meaningless for defining the concept "homology." Boyden (1947) says: "Today the pendulum has swung so far from the original implication in homology that some recommend that we define homology as any similarity due to common ancestry, as though we could know the ancestry independently of the analysis of similarities!" As though it mattered for the definition of the concept "truth" that we cannot recognize truth itself, and everywhere in science are limited to erecting hypotheses concerning truth.

Naturally, in determining homologies we are limited to erecting hypotheses— such as that particular characters a, a', a" belong to a phylogenetic transformation series. We produce this hypothesis with the aid of a comparison of different organisms, which is an attempt to determine correspondences between their individual characters. "In comparing two things we set up a one-to-one relation or correspondence between the parts of the one and those of the other and proceed to state how corresponding parts resemble or differ from one another with respect to certain sets of properties" (Woodger, cited from Cain and Harrison 1958). Determination of morphological correspondences consists, according to Woodger, in choosing "a pairing which will bring the maximum number of parts into identity correspondence with respect to certain of the morphological properties." Woodger's definitions correspond to a theory of comparison worked out by symbolic logic, and to that extent are exact. In the more graphic words of Remane (1952), homologous characters (or better, parts of organisms) must conform to "the criterion of sameness of position in comparable fabric systems." This is one of Remane's three "principal criteria of homology." But with respect to defining the concept "homology," all three of his "principal criteria" are only accessory criteria that we have to use because the real principal criterion—the belonging of the characters to a phylogenetic transformation series—cannot be directly determined. Remane's second and third principal criteria are in reality only accessory criteria of lower rank that are not usable without the "criterion of sameness of position."

Intermediate forms that facilitate making up morphological correspondences (Remane's "criterion of the linkage of intermediate forms") can be inferred from the ontogeny of the organism stages originally compared, or they may be supplied by paleontology. It should be strongly emphasized that such "intermediate forms" are intermediate not only with respect to the characters whose homology is to be investigated, but that their overall plan makes it possible to bring an even greater number of parts into "identity correspondence." Remane's third principal criterion, the "criterion of special quality of the structures," is, like his first, included in Woodger's definition.

Finally, the concepts of symplesiomorphy and synapomorphy go somewhat beyond the range of what are ordinarily called "homologous characters." We started from the idea that a, a', a" are different characters in a transformation series. We can speak without reservation of homologous characters if a, a', a" are transformation stages of an organ. But the transformation a–a'–a" may also consist in the complete reduction of the organ. For example, the absence of the wings in fleas is undoubtedly an apomorphous character in comparison with the presence of wings in other holometabolic insects. On the other hand, the possession of wings is an apomorphous character in comparison to their absence in the so-called "Apterygota." In general we speak only of the homology of organs, but a "character" may also be the absence of an organ. This discrepancy between the concepts "organ" and "character" explains the tortured impression produced by many phylogenetic discussions that try to make do with concepts such as "special homology," "limited homology," and so on (instead of "synapomorphy"). It is completely unequivocal to say that the absence of wings in the Anoplura and Mallophaga is a synapomorphous character, whereas in the Collembola, Protura, etc. it is a symplesiomorphous character. This cannot be expressed in an equally unequivocal way by saying that the absence of wings is a "special homology" in the Anoplura and Mallophaga, but not in the Collembola, Protura, etc.

(2) *Character phylogeny.*—If it can be shown that a character is homologous in a series of species, the question arises: In which direction is this transformation series to be read? In other words, what is the plesiomorphous ("primitive extreme" of Maslin), and what the most strongly apomorphous ("derived extreme" of Maslin) character condition? Research aimed at answering this question is called "character phylogeny" by W. Zimmermann. Maslin (1952), whose arguments we will occasionally follow, calls a transformation series a "morphocline."

Since we do not know from direct observation the direction in which a transformation series took place, we are dependent on accessory criteria here too. These are:

(a) Criterion of geological character precedence. If in a monophyletic group a particular character condition occurs only in older fossils, and another only in younger fossils, then obviously the former is the plesiomorphous and the latter the apomorphous condition of a morphocline a, a′, a″. We will postpone a more detailed discussion of this criterion until a later section dealing with the paleontological methods of phylogenetic systematics (p. 142).

(b) Criterion of chorological progression. "Each step in phylogeny consists of a shift of an organism toward an equilibrium with a changing environment." (Maslin 1952). If this is correct, it must be assumed that when species A splits into two successor species B and C (Fig. 21, 1), the transformation of a character a into a′ will appear in the species C that has departed farther geographically or ecologically from the initial species A. Thus there would be a certain relation between "apochory" and apomorphy that can be used to determine the direction of the transformation. We will also postpone answering this question until a later section (p. 134).

(c) Criterion of ontogenetic character precedence (Naef). This assumes that the transformation of a character during ontogeny "recapitulates" the phylogenetic transformation of this character, so that the direction of the transformation from plesiomorphous to apomorphous condition during phylogeny can be determined from the sequence of ontogenetic stages. This criterion belongs in the sphere of ideas of the biogenetic law. In the same sense, Naef speaks of a primacy of ontogenetic precedence, and formulates his "law of terminal deviation" accordingly: "In the course of phylogenetic change, the stages of a morphogenesis are the more conservative the earlier they occur in the ontogenetic series, and the more advanced the later they occur." Zimmermann (1953) calls Haeckel the true founder of the law of recapitulation.

Sewertzoff (1931), in particular, has examined the relations between ontogeny and phylogeny in animals ("phylembryogenesis," in his terminology). "Phylogenetic changes in the characters of adult animals (form, structure, size, position, etc.) take place in progressive evolution in three ways: (1) by addition of end stages, or anaboly, (2) by deviation in the middle stages of morphogenesis, and (3) by change of the first anlagen, or archallaxis." Sewertzoff's investigations were aimed particularly at determining the extent of validity of the biogenetic law. He concluded that the "law of recapitulation of Müller-Haeckel" is completely valid in developmental series of the first type (anaboly), whereas in the second type only the embryonic characters of the ancestors are repeated, and the law does not hold at all in the third type.

The investigations of Sewertzoff and others have concerned not only the extent to which the ontogenetic transformation of organs or characters "recapitulates" the course of transformation of the same characters in phylogeny. They have also tried to determine whether the relative velocities of different series of transformations (a, a', a"; b, b', b") during ontogeny permit a conclusion about the relative velocities of these transformations in phylogeny. If, for example, the character condition b" is reached simultaneously with a', can we assume that this was likewise true during phylogeny? The investigations have shown (as could probably have been predicted as most likely) that important shifts in the simultaneous appearance of particular transformation stages of different characters can result from retardation or acceleration of individual developmental processes (transformations) during ontogeny. For phylogenetic systematics this means that the transformation stages that led to a character condition during phylogeny, as it is realized in the adults of a particular species, cannot be read off with certainty from the ontogeny of the species. Nevertheless the "criterion of ontogenetic character precedence" remains an important aid in phylogenetic systematics, provided it is not uncritically evaluated more highly than other aids that, under certain circumstances, may lead to different conclusions. Complete rejection of the law of recapitulation (de Beer 1959) is certainly unjustified.

(d) Criterion of the correlation of series of transformations. If the individual stages of several series of transformations usually or always appear together in different species or species groups (a with b, a' with b', a" with b"), the characters are said to be correlated. Such correlations are of significance to phylogenetic systematics only if the direction in which one of two or more correlated trans-

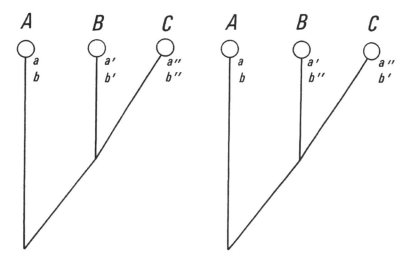

Figure 24. Correlation of several-stage transformation series of two characters. If for the series a, a', a'' it is known for sure that a represents the plesiomorphous and b'' the most apomorphous condition, then also in the series b, b', b'' only b can represent the plesiomorphous and b'' the most apomorphous condition. In the mode of division shown at the right the monophyly of B follows from this, though its occurrence could not be demonstrated with the aid of the transformation series a.

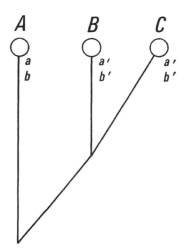

Figure 25. Correlation of two-stage transformation series. If for the transformation series a, a' it is known for sure that a represents the plesiomorphous and a' the apomorphous condition, then for the series b, b' it cannot be concluded from this that b represents the plesiomorphous and b' the apomorphous condition or vice versa.

formation series is to be "read" is known. Under certain circumstances this knowledge gives the direction of the other transformation series. As Fig. 24 shows, if the direction of the morphocline a, a', a" is known (a is plesiomorphous, a" most strongly apomorphous), the conclusion that the species B and C are most closely

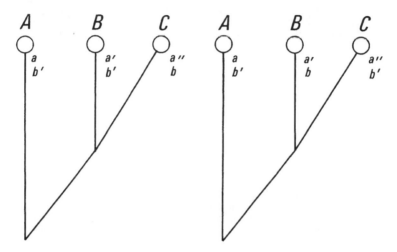

Figure 26. Correlation of two transformation series, one of which (a, a', a'') is more richly subdivided than the other (b, b'). If it is certain that in the a-series a represents the plesiomorphous and a'' the most apomorphous condition, then in the b-series only b' represents the plesiomorphous and b the apomorphous condition. In a division as in the right diagram, b can establish that the group in which this condition occurs (B) is monophyletic, which would not be possible on the basis of character a' alone.

related is compatible only with the assumption that in the transformation series b, b', b" the plesiomorphous condition is represented by b and the most strongly apomorphous condition by b".

This is not true if the correlated transformation series consist of only two distinguishable stages (a, a' and b, b'). Then (Fig. 25) if condition a is plesiomorphous, the assumption that species A and B are closely related is compatible under certain circumstances with the interpretation that b is plesiomorphous as well as that b' is plesiomorphous. Consequently such cases of correlation are of no significance to systematics. In a different kind of character distribution (Fig. 26), an interpretation of the series b is possible. If one of the series of transformations is more richly differentiated than the other (for example, a, a', a" and b, b'), the correlation must necessarily be incomplete: b' may occur together with both a and a', and other combinations are possible. Systematics can profit in cases where the more richly differentiated series is interpreted with the aid of the simpler series. Thus in Fig. 27 the more simply differentiated series a, a' per-

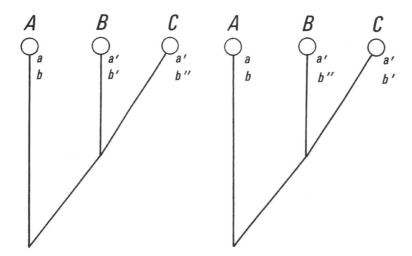

Figure 27. Correlation of two transformation series, one of which is more richly subdivided than the other. If it is certain that a represents the plesiomorphous and a' the apomorphous condition, then for the b-series it can only be true that b represents the plesiomorphous and b'' the most apomorphous condition. The occurrence of the transformation stage b'' can then in each case prove the group in question (C in the left diagram, B in the right) to be monophyletic, which would not be possible with the aid of the a-series alone.

mits the recognition that B and C form a monophyletic group. But the fact that, within the group BC, C represents a monophyletic (partial) group would not have been recognizable with the aid of a, a' alone; it is possible only with the aid of b, b', b''.

Two special cases of correlation in transformation series (morphoclines) are described by Maslin (1952): "If one extreme of a morphocline resembles a condition found in the less modified members of related groups of the same rank, this extreme is primitive," and "If two or more morphoclines appear in different groups of organisms, and an extreme of morphocline occurs in the same taxonomic unit, then these extremes are the primitive extremes."

The formulation of these two special cases, the first of which was already called the "principle of systematic character precedence" by Naef, is perhaps not unequivocal. There are, however, relatively simple considerations that would allow us to derive many special cases of correlations between different transformation series as deductions from the basic assumptions. Many of these could certainly be illustrated with concrete examples. We will forego doing so, especially since the discussions that would have to accompany them relate rather to the question of how systematics deals with the limitations inherent in all its accessory criteria for discovering phylogenetic kinship relationships. We will have to try to answer this question later anyway.

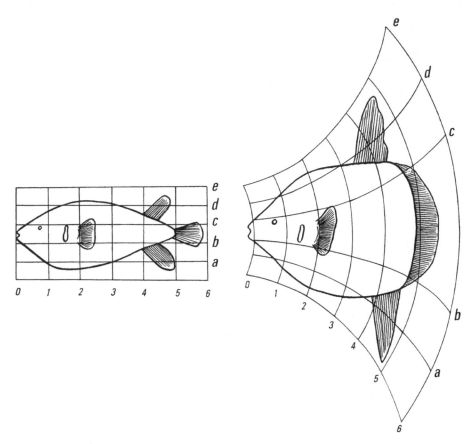

Figure 28. Transformation of the body outline of the porcupine fish (**Diodon**, left) into that of the sunfish (**Orthagoriscus**, right) through transformation coordinates. (Drawn after Thompson.)

Only one special form of correlation between different transformation series will be discussed briefly: allometry (Fig. 28). According to von Bertalanffy the law of allometry "directly controls some of the most important phylogenetic series." In positive allometry, "the larger the body becomes, the larger does the allometrically growing organ become, not only absolutely but also relative to body size. With very small body size only traces of the organ are present; the larger the body becomes the more prominent is the organ. Any factor that increases body size must result in enlargement of the organ, not only absolutely but also relative to body size." This law makes it possible to reduce many apparently complex differences in form to very simple bases: "The polymorphism of worker castes in ants can be explained as due to differences in the time of pupation. The small workers (with relatively small heads) are individuals that were induced to early pupation by the ant nurses, whereas the larger soldiers with immense heads are individuals with the same hereditary background that were allowed to grow to maximum larval size, so that they not only attained absolutely greater body size but, following the allometric principle, the head became relatively particularly large" (von Bertalanffy, II).

The relatively more pronounced expression of certain characters in larger individuals of a species is not limited to measurable characters in the usual sense. For example, among insects such larger individuals are often more strongly bristled (greater numbers of bristles and stronger expression of them, e.g., on the legs; presence of bristles that are completely absent in smaller individuals), and certain colors (e.g., red) relatively much more extensive than in smaller individuals. The males of many species of *Sepsis* are an example. This phenomenon has led many systematists not familiar with it into regarding large individuals as new species because they are not simply enlarged editions of normal individuals.

The formula for allometric growth is $y=bx^a$ (or $\log y = \log b + a \log x$), where x is body size, y is the allometric structure, and a and b are constants. Therefore the similarity between several forms that differ only allometrically (as compared with the differences from other groups that do not belong to the same series) is expressed by the constants a and b. Of course, the laws of allometry are only a special case of a (functional) correlation between two or more characters, however important and interesting they are. At least in morphology, we usually think of a correlation between two characters as linear, whereas allometry is a logarithmic (although still relatively simple) correlation. This is of no importance for phylogenetic systematics, because in determining the direction in which transformation series are to be read it makes no difference whether the characters are linked in linear or allometric correlation. Knowledge of allometric correlation is important, however, because it permits recognition of linkages between different series of transformations that might otherwise be unrecognizable.

In discussing the "holomorphological" methods of phylogenetic systematics we have so far considered only morphological characters in the narrow sense. But it is an old demand that systematic work should consider "all characters" as far as possible, and not be limited to only one sector of the fabric of characters. Consequently other characters not narrowly morphological have recently been recommended to an increasing degree. Insofar as these are intended to supplement and test the results based on morphological characters, phylogenetic systematics will profit from them. A prerequisite for this is that nonmorphological characters be used according to the same strict, theoretically incontestably based methodological principles as morphological characters are.

Unfortunately this is rarely done. In works dealing with the systematic evaluation of nonmorphological characters it is often impossible to tell whether the author intends to promote the recognition of typological ("affinity" in the sense of overall resemblance) or phylogenetic relations. Such investigations are useless for phylogenetic systematics. Erdmann (1927) recommends determining "kinship relationships" of organisms with the aid of transplants. She starts out from Leo Loeb's concept of the individual differential. "The sum of all differences—be they chemical or physical, known or unknown, in short the sum of all differences between this individual and others of the same species—are defined as the individual differential. In the same way we will define specific and 'stem' differentials." "This individual differential is biologically demonstrable through the reactions between the living cells and those of another individual of the same species. Usually these reactions have been investigated only in hybridization experiments." "The surest proof of the existence of an individual differential is by the method of transplantation." "The different lengths of time required by pieces of skin grown in vitro to heal into a foreign environment into which they are later implanted" is a measure of "kinship relations." Here kinship relations clearly means typological relationship.

Other authors, with much more modest aims, recommend supplementing morphological studies with chemical investigations: "All life processes can be reduced to chemical and physicochemical reactions, and it seems completely certain that there are as many types of metabolism as the inconceivable number of forms of organisms. Moreover, there is not only a species chemistry; the various organs, and even tissues and cells, have a specific metabolism. This opens the possibility that investigation of metabolism, rather than study of the often extremely similar morphological characters, may disclose characteristic differences, or that only by these means can the kinship relations of aberrant groups be revealed" (Krüger 1930).

"Metabolic types" (as well as all other "chemical characters" of the species) can be used for disclosing phylogenetic relationships only if series of transformations—similar to those among the morphological characters—can be recognized among them, and if plesiomorphous conditions can be distinguished from apomorphous conditions in these series. Certain observations indicate that it is in fact possible to distinguish between plesiomorphous and apomorphous conditions, at least in gross terms, in the chemistry of species.

Blagoveschenski (1929) has developed detailed ideas concerning phylogenetic hysteresis. He starts out from the peculiar distribution of alkaloids in the plant kingdom. These are found on the one hand in the terminal families of the phylogenetic series, and on the other hand in forms that retain archaic and primitive characters that reveal the early completion of a phylogenetic series. Something similar is true for genera within families in which not all genera have alkaloids. Alkaloids are highly stable aromatic compounds, and therefore their accumulation is a sign of chemical senility, of "phylogenetic hysteresis." The accumulation of great numbers of such stable compounds reduces the reaction capacity of protoplasm, and this in turn limits the possible paths of development—just as according to Pictet the death of protoplasm is connected with the stabilization of protein molecules by the formation of ring compounds. Thus there are "chemically old" and "chemically young" groups, characterized by the quantity of cyclical compounds. For this reason the variability of organisms was probably greater in earlier geological periods than it is today, because the once relatively unstable systems were transformed by manifold impulses from the environment into more stable systems. In connection with the accumulation of stable compounds, lability had to decrease, to the point of cessation and extinction of the species (Rosa's rule of progressive reduction of variability). In this way, according to Blagoveschenski, evolution tends toward the reduction of free energy, i.e., toward the most probable equilibrium. In every phylogenetic series one can distinguish stages of youth, maturity, and senility, with an accompanying decrease in plasticity [von Bertalanffy, II].

For such observations to be useful to phylogenetic systematics, investigations cannot be limited to random samples scattered over the whole animal kingdom. Actually all species (at least of a particular group of animals) should be considered. Blagoveschenski (1929) himself points out that even such complex chemical compounds as vanillin and caffein occur in very widely separated groups, and must have originated independently several times. According to his data, vanillin occurs in *Vanilla, Avena, Lupinus, Pinus,* and *Beta*; caffein in *Coffea, Thea, Ilex, Cola,* and *Theobroma.*

Thus there is convergence even in the realm of "chemical characters." But where convergences occur or are to be suspected, only the most subtle distinction of the individual characters and the most discriminating evaluation will protect us from false conclusions (see p. 117). When Kutscher and Ackermann (1926) state that 100 to 200 kilograms of animals are needed for each investigation of "animal alkaloids," which they propose as a "supplement to morphological systematics," this alone is sufficient reason for doubting that this method will ever be of great significance for systematics.

Failure to understand the limits of the usability of chemical characters in phylogenetic systematics leads to completely grotesque overevaluation of the results of investigation. Schmalfuss and Werner (1926, 1930), for example, investigated pigment formation in the animal organism. On the basis of their results, the animals investigated were divided into two groups:

"1. Species whose hemolymph yields a heat resistant and a heat nonresistant component, which changes $1-\beta-3,4-$dioxyphenyl$+$aminopropionic acid into pigment: *Agrostis pronuba* L., *Cimbex variabilis* Klug, *Tenebrio molitor* F., *Tegenaria domestica* L.

"2. Species from which only the heat resistant portion could be obtained: *Aeschna* (*cyanea* Mull.), larvae, *Rhabdophaga salicis* Schrank, larvae, *Cancer pagurus* L., *Ceratophyllus canis* Curt., *Homo sapiens.*"

Far-reaching phylogenetic conclusions are drawn from this table: "This finding supports the view that the fleas are closer to the Diptera than to the beetles, because within the families the ferments react similarly" (1926). According to Schmalfuss, the findings support the "complete separation of the crustaceans from the arachnoids" because the two representatives that were investigated are listed separately in the two groups and not united in one group. What would one say of a phylogenetist who based his conclusions as to the phylogenetic relationships between two classes on a morphological comparison of two species?

Albumen plays a special role in the attempts to use chemical characters for determining kinship relations. This is probably because, as the "bearer of life," among other things, to many authors it also seems to a special degree to be the "bearer of kinship relationships."

The oldest and most widely used method that attempts to determine relationships from differences in species-albumens is serology. "The use of this method is based on the fact that species-specific albumens of two species are the more similar the more closely the animals are related to one another. This indicates that new units of albumen structure appear with every morphological or physiological change in a species or race, which are probably also the causes of every external phenomenon of life. In order to have a short name for such units we will call them proteals" (Mollison 1924). On the basis of the evidence of these proteals it is said to be possible to determine not only the degree of relationship but even the relative level of organization of the species being compared:

Equal quantities of human serum and human antiserum are mixed, and equal quantities of macaque serum once with macaque antiserum and once with human antiserum (all in isodatic amounts). The quantities of precipitate of the heterologous reaction are expressed as percentages of the homologous reaction. The human antiserum in macaque serum produces a much heavier precipitate in relation to the homologous reaction than does the reverse mixture, macaque antiserum in human serum. Apparently this is explained by the fact that the occurrence of new proteals in man, the more highly differentiated species, reduces the quantity of the primitive primate proteals (which the macaque also possesses), whereas the primitive proteals still occupy a large volume in the precipitogen of the macaque. Reciprocal reactions between chimpanzee and macaque are similar. Therefore the species closer to the common stem of the two has the proteals common to both species in greater quantity [Mollison, according to Karny 1925].

In addition to this "precipitin method," Zimmermann (1937) distinguishes the "anaphylaxis method," the "agglutination method," and the "complement binding method," all of which are based on the same assumptions as the precipitin method but are not as highly perfected.

The great prestige of serum diagnostics among many biologists, particularly as applied to plants (Mez), vertebrates (Mollison 1924, Johnson and Wicks 1959), invertebrates (crustaceans: von Dungern; insects: O'Rourke 1958), is probably based mainly on the fact that its relatively easily gained results in most cases actually agree with the results of analysis of morphological similarities, which are much more difficult to achieve. Consequently there is a temptation to rely on it even where the results of the usual comparative morphology are uncertain and several interpretations are possible. Serum diagnostics always gives an unequivocal result in a way that suggests the exactness of laboratory experiments, which to many is the acme of scientific research. Mez (1924), the creator of the serodiagnostic "Königsberg plant phylogeny," contrasts experimental systematics working with serodiagnostics with taxonomy working with other methods, which he calls "transcendental systematics." Such ideas were properly criticized by Tschulok (1910): "Calling serological agreement 'experimental proof of descent' is a gross logical error" because "experiments work with real relations, whereas systematic relationship is based on ideal relations." To speak of an experimental systematics is a gross logical error whether the system based on sero-diagnostics is regarded as phylogenetic or typological, since in either case systematics determines relationships between organisms that in this sense are ideal, whereas experimentation determines action relations ("real" in this sense). Making albumen similarity visible by the precipitin method is thus equivalent to making optical-morphological structure visible by means of light filters or a mounting medium that has particular refraction characteristics.

The decision as to the importance of serum diagnostics in phylogenetic taxonomy depends on whether there is actually a firm, unequivocal relation between degree of phylogenetic kinship and "albumen relationship." In other words, do two species that arose from a parent species that is common only to them always have a greater albumen relationship than two species each of which goes back, with a third or fourth species, to an even more recent parent species? Mollison himself says (see above) that the results of serological tests differ depending on which of two species being compared is selected as the serum donor, and that the species closer to the common stem (i.e., the one that is more similar to the common stem species) "has a greater quantity of the common proteals" than the other. This clearly answers the question in the negative, which simply means that serum diagnostics produces a typological system in the sense of Fig. 24 (right). Consequently albumen relationship is a "gestalt relationship" (or "form relationship," as it is more commonly called, although in this case not quite correctly), and not a phylogenetic relationship. That it is more important than similarity between species in any other character (Mollison 1924, von Krogh 1938) even in typological systematics could actually be proved only by showing that albumen is "the root of all morphological characters," and that "all likenesses would have to be explained by agreements in albumen substances, just as all differences are explained by albumen differences (Mez 1926). In that case one could, as Karny (1925) hoped, actually "express each systematic category of organisms in terms of chemical differences in certain molecules or groups of molecules, once biochemistry has advanced far enough." Serum diagnostics could be of great importance in phylogenetic systematics only if transformation series like those we have already discussed for morphological characters could be erected for albumen differences. We would also have to be able to distinguish between plesiomorphous and apomorphous conditions within these transformation series. Only then could we ask whether synapomorphy in albumen structure is an expression of all gestalt characters of the species concerned. Only in this case could serum diagnostics replace all other methods in phylogenetic systematics, because it would not share their limitations.

It is extremely unlikely that serum diagnostics could ever assume this importance. Even Mollison (1937, 1938) admits that the method still works only rather coarsely. The domestic buffalo and domestic cow, or goat and sheep, still cannot be differentiated serologically. The clarity of its results is also impaired by the fact that, in part, different organs give different reactions. In a comparison of rye, wheat, and barley, rye pollen serum gives reactions only with the seeds of wheat and barley, not with other organs (Karny 1925). As of now there is not a chance of fulfilling the other requirements that we posed above. At present, serodiagnostic findings express in any case only a kind of "form relationship" of organisms: "The serological data actually result in a multidimensional expression" (Boyden 1959). Zimmermann's question (1937) of whether serology has any justification at all beside other (much better worked-out and theoretically based) methods can thus relate at the moment only to typological systematics.

Paper chromatography (paper electrophoresis) has recently been used to determine differences in albumen structure. Stephen (1958), in a review of the most important studies, summarizes the most important results: "Extrapolation from these studies tends to indicate that comparative protein analysis not only distinguishes large categories in a qualitative manner, but could be used to differentiate between populations (and individuals) of polytypic forms." Dessauer and Fox (1956) have even erected a "tentative ordinal key to the Amphibia and Reptilia based on the plasma proteins." Regarding this method, nothing more,

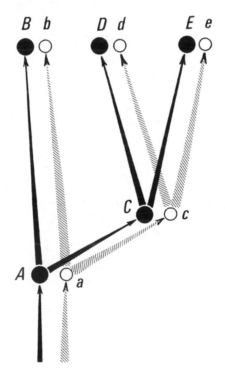

Figure 29. "Fahrenholz' rule." Complete parallelism of speciation in a host (black, capital letters) and a parasite (white, small letters) group. Each host species corresponds to a parasitic species; the most closely related hosts (D, E) have the most closely related parasites (d, e).

and as far as its usefulness in phylogenetic systematics is concerned nothing more favorable, can be said than for serum diagnostics and the previously discussed chemical methods.

In a sense related to serum diagnostics is a method that attempts to infer relationships of host species from the proven relationships of their animal parasites. Eichler (1941) formulated the assumption that "in the case of permanent parasites . . . the relationship of the host can usually be inferred directly from the systematics of the parasites" as "Fahrenholz' rule." "Paralleling the evolution and splitting up of the hosts" there is supposed to be a corresponding evolution and splitting up of the parasites (Fig. 29). Eichler's "rule of development," which says that "among equivalent larger systematic units of hosts, the most differentiated groups have a greater variety of (mostly permanent) parasites than do less differentiated groups," scarcely needs to be regarded as an independent rule since it follows from Fahrenholz' rule.

There is a certain irony in the fact that Fahrenholz, on whom Eichler bases his ideas, speaks of the "blood relationship" of the host, which the occurrence of the same or similar parasites is supposed to prove. It is decisive, however, that by "blood relationship" he does not mean a phylogenetic relationship in the true sense, but a "juice relationship" or similarity in the material construction of organisms of the same kind as serum diagnostics attempts to find. Hering (1926) clearly expressed the idea that the frequent restriction of parasites to a particular narrow group of host organisms (their oligophagy) is based on the same assumptions as the albumen relationships established by serum diagnostics. Hering (1955) also noted that in evaluating this method it is relatively unimportant whether the host species are animals or plants: the attempt to infer relationships among host plants from the relationships of their phytophages "would mean an application of the Fahrenholz rule, which was derived from lice living on animals, to the phytophages."

Unfortunately in almost none of the works dealing with this group of questions is there a clear distinction between the concept "typological relationship" = affinity in the sense of Cain and Harrison (1958), and "phylogenetic relationship." In this respect the authors of the most modern studies (e.g., Stammer 1957) have not gone beyond Fahrenholz.

In his investigation of leaf-mining insects, Hering came to essentially the same conclusions as those of the "Königsberg phylogenetic tree," which were based on serum diagnostics. But we must remember that the relationships determined by serum diagnostics are typological and not primarily phylogenetic. We must also remember this in evaluating a comprehensive judgment of Hering (1955) that in the vast majority of cases the choice of food of insects (except in completely polyphagous species) is determined by the kinship relations of the plants, but that this principle is obscured by a great number of conditions that can decisively alter the choice.

Hering gives many examples showing that the "choice of food" of phytophagous insects is not determined only by the "relationships" of the host plants (i.e., not even by typological "relationships," which at least in many cases actually correspond to phylogenetic relationships). Stammer (1957) gives numerous examples proving that the relationships between animal parasites and their host species cannot simply be brought under the denominator "parallel phylogeny" either. The most varied forms of similarity relationships, from similarity of station or environment (even where this is expressed only in the hosts' belonging to a common sleeping community, as in bats), on up to the construction of the body from chemically similar substances, may be responsible for the occurrence of related parasites.

Scarcely anyone seems to consider, however, that even in parallel phylogenies between hosts and their parasites there can be cases in which the host-parasite linkage does not reflect phylogenetic relationships even though there has been no transmigration of parasites from one host to another.

Assume (Fig. 30) that a species A was colonized by a parasitic species a. The host species then divided into the species B and C, and at the same time the para-

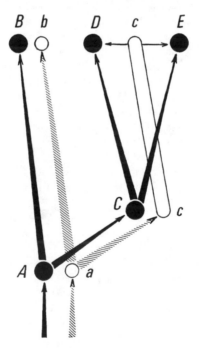

Figure 30. "Fahrenholz' rule." Incomplete parallelism of speciation in a host (black, capital letters) and a parasite (white, small letters) group. The parasitic species c has not participated in the splitting of the host species C. Its occurrence in the host species D and E corresponds to the fact that these two form a monophyletic group.

sitic species divided into b and c. Later the host species C divided into species D and E, but its parasitic species c did not divide. In this case the occurrence of the same parasitic species c in the two host species C and D truly corresponds

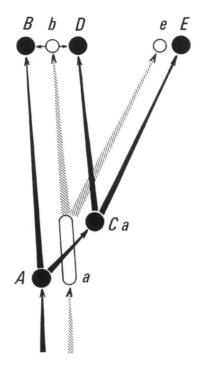

Figure 31. "Fahrenholz' rule." Incomplete parallelism in a host (black, capital letters) and a parasitic (white, small letters) group. The parasitic species a has not participated in the splitting of the host species A. In the splitting of the host species C into the species D and E, however, a new parasitic species e has arisen on E. The host species B and D, colonized by the parasitic species b, do not form a monophyletic group.

to the closer kinship of these two species. But now the reverse case (Fig. 31): the parasitic species a did not divide when the host species A divided into B and C; it still colonizes the two daughter species B and C as a single species. Not until the host species C divides is there also a division of a into two daughter species. But of these two daughter species only one (e), on the host species E, becomes morphologically different from the stem species a; the other (b), living on D, has not changed. In this case the occurrence of the same parasitic species b (with at least morphologically identical populations) on the two host species B and D does not reflect closer phylogenetic relationship; D is phylogenetically less closely related to B, with which it shares the parasitic species b, than with E, which has a different parasitic species (e)!

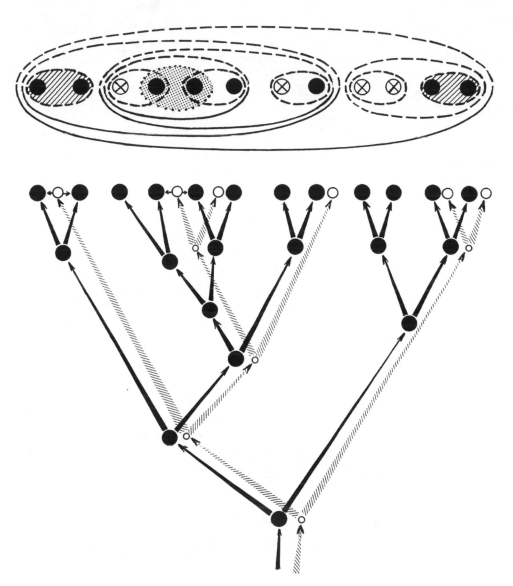

Figure 32. "Fahrenholz' rule." If parasitic groups do not participate in individual speciation processes of the host group, the phylogenetic kinship relations of the host species are only very imperfectly reflected by the division of the parasites (even if migration from one host to another has not taken place). In the upper diagram monophyletic host groups, which are also recognizable from the distribution of the parasites, are completely outlined and shaded with diagonal lines. Monophyletic host groups that are also recognizable from the distribution of parasites, if one neglects the fact that some species harbor no parasites (circles with crosses), are outlined with a solid line below and a dashed line above. A group that could falsely seem to be monophyletic because of the occurrence of identical parasites is dotted.

This example shows that we cannot infer with certainty from the occurrence of one and the same parasitic species in two different host species (or species groups) that the latter form a monophyletic group in the sense of our definition (p. 73), even though there has been no transmigration from one host to the other.

In Fig. 32 this is shown for an even more richly differentiated host group. In this example we make the further assumption that, with the splitting of one of the host species, the parasites were not taken over by both daughter species— either the starting species from which the parasite-free daughter species arose was accidentally not parasitized, or the parasites subsequently perished. This example also shows that even with parallel evolution of hosts and parasites (in the sense of Fahrenholz' rule) a monophyletic parasite group does not necessarily correspond to a monophyletic host group.

Even the most extreme advocates of the thesis that the phylogeny of the parasites usually parallels the phylogeny of the host (in the sense of Fahrenholz' rule) do not assume that the parallelism is so close that every process of speciation in the one corresponds to a process of speciation in the other. Stammer (1957) states explicitly that "it is a characteristic of almost all parasites for which it is possible to make comparisons that their phylogenetic development is retarded and progresses more slowly than that of their hosts. The reason for this retardation lies, as Szidat (1956) correctly points out, in the constant environment in which the parasites live."

If this is correct, then there is no guarantee whatever that we can infer that the host species form a monophyletic group from the presence in them of a certain species of parasite or a monophyletic group of parasites. Two examples will illuminate the significance of this fact:

Eichler (1948) writes:

We encounter even more drastic examples demonstrating such parallel evolution of parasite and host in cases where the different parasites belong not only to different genera or families, but even orders or phyla. This is the case in the ostrich and rhea, for example, where not only the Mallophaga, but also two equivalent mites (*Pterolichus bicaudatus* and *Paralges pachycnemis*) as well as similar cestodes and nematodes show that we are dealing with natural relatives. Thus the two southern continents have not each evolved their own giant flightless bird, but the ostrich and rhea are relatives that were isolated on different continents.

Stammer (1955) refers particularly to the investigations of his pupil Rühm: "up to eleven species of nematodes (in *Hylastes ater*) could be identified in one ipid species. With this abundance of species, Rühm was able to discuss the relationships of the individual species, genera, tribes, and subfamilies of ipids." The following statements show how Rühm (1955) argues: "If we consider, for example, the nematodes of *Hylastes ater* and those of *Hylastes cunicularis* we must recognize a complete morphological identity of the beetles so far known, and also an equivalent faunal composition. The species that correspond to one another, *Parasitorhabditis ateri* (Fuchs 1937) Rühm 1955 for example, are also bound to their hosts in the same way, namely as commensals (terminal larvae!)." "On the basis of the nematode fauna it is to be concluded that the separation of the two species of *Hylastes*, for example, must have taken place very late, and that because of equivalent biological conditions for the nematodes the same faunal composition is also preserved. There is no doubt that the hosts are closely related to one another."

On the contrary, we could show (Figs. 29-31) that neither the possession of identical, nor of the most closely related parasites, permits the certain conclusion that the hosts are more closely related to each other than to other species on which the parasites in question do not occur. Exactly the same applies to the evaluation of Eichler's statements on the relationship of Ethiopian and Neotropical ratites.

"Mayr has called . . . parasitology a gold mine for evolutionists and biologists. One can emphatically confirm these words" (Stammer 1957). If this is to apply to parasitology with respect to phylogenetic systematics, the theory of the parasitological method must be much better worked out than it is now. For phylogenetic systematics it is not so important to know that host species harboring identical parasites or a monophyletic group of parasites are "natural relatives" (Eichler) or "closely related" (Rühm). Rather it is a question of the degree of phylogenetic relationship, of deciding whether the host species are more closely related to one another than to species on which the parasites in question do not occur. Consequently, as in the morphological method, there must be criteria for determining when this assumption is justified (Fig. 30, c) and when it is not (Fig. 31, b). No such criteria are known.

The following two examples show wherein, in the present unsatisfactory state of its theoretical basis, the importance of the "parasitological method" for the phylogenetic method lies. "One of the most beautiful and best analyzed examples is furnished by the flamingo, which even modern ornithologists regard as only an 'aberrant stork' (whose ducklike bill is only a phenomenon of convergence). Now, four different species of Mallophaga are known from the flamingo. . . . Three of the four bird lice of the flamingo correspond completely and entirely to the mallophagans of the ducklike birds, and show no close relationships to those of the storks. And even the fourth mallophagan of the flamingo speaks for a relationship in the direction of the ducks" (Eichler 1948).

This corresponds exactly to an example that Hering (1955) called a classical case, in which the phylogenetic relationships of oligophagous phytophages indicate the phylogenetic relationship of the host plant that had not been recognized at first. "The plant genus *Brunfelsia*, placed by contemporary botanists in the Scrophulariaceae, is the food plant of a *Thyridia* caterpillar (Lepidoptera, Neotropidae). All neotropids live exclusively on Solonaceae, in which family *Brunfelsia* was later correctly placed."

In both of these cases a host had been incorrectly allocated in the system far outside the monophyletic group to which it belongs as a partial group. In spite of its relationships to serum diagnostics and the use of chemical characters in systematics, the parasitological method points in many respects beyond the holomorphological method. It has distinct points of contact with the chorological method (p. 133), and in many respects it is much more important for determining the absolute rank of systematic categories than are the strictly holomorphological methods.

On the other hand, attempts to use modes of behavior in phylogenetic systematics clearly belong in the field of holomorphological methods. "The establishment of relationships and of phylogenetic systems with the aid of behavioral characteristics implies studies of comparative behavior just as the use of morphological characters for this purpose implies studies of comparative morphology" (Michener 1953). This is completely correct, but presupposes that series of transformations, similar to those for the morphological characters in the narrow sense, can be recognized for behavioral characters. "It seems obvious . . . that the idea of homology is useful and necessary in connection with behavior" (Michener 1953). But beyond this it must be possible to distinguish between plesiomorphous and apomorphous conditions in the series of transformations. As with true morphological characters, only correspondence in "synapomorphous" modes of behavior can prove that different species belong to a monophyletic group.

Unfortunately behavioral science does not yet follow this principle. This is clearly evident in the statements by Meidmann on "the systematic value of courtship behavior in *Drosophila*." In contrast to Sturtevant, Weidmann concludes that in *Drosophila* "behavioral elements are indeed useful systematically, provided one knows thoroughly the species concerned." He describes his method as follows: "Following the procedure of Lorenz (1941) and Sturtevant (1942), the number of actions in which two species differ is counted, i.e., all actions that appear in one species but not in another are added. Actions that are common to both species, or which are absent in both, are not considered. The smaller this number is, the closer the kinship between the two species, whereas a great number of differences indicates a greater systematic separation." The "relationship" that can be determined by this method is typological, not phylogenetic!

Very interesting beginnings for the use of modes of behavior in phylogenetic systematics are contained in the statements of Kessel (1955) on the mating activities of balloon flies. Kessel distinguished eight evolutionary stages:

Stage 1. The first stage is simply that of a predaceous species in which the male does not bear a wedding gift for his bride. . . . The prey is captured and devoured by the sexes independently. This stage differs in no way from the cases of ordinary predaceous species, such as those which occur in other insect families.

Stage 2. In the second stage the male avoids any cannibalistic attention on the part of the female by carrying with him, ready for presentation at the moment they embrace, a wedding present in the form of a juicy insect. The male first catches the prey, and then he flies about with it in search of a mate. Often the prey carried by the bride-hunting male is as large as he is. . . . In this second stage the prey is simply eaten by the female, the gift serving merely to occupy her attention as she passively accepts the male during the time she is thus distracted. The mating lasts as long as the food does.

Stage 3. While the third stage is similar to the second, it differs in two important respects. As in the second, the male first catches the prey, but in this stage he does not go in search of a female. Instead, he joins other males, bearing gifts, in an aerial dance of eligible bachelors. When a female approaches, the males do not take after her, but instead continue to dance in the swarm. The female, on approaching such a swarm, must take the initiative. In this third stage, the prey has become the stimulus for mating, and it no longer serves merely as a distraction to save the male's life. The female enters the swarm and selects one of the dancing males, they embrace and she accepts the gift, and the two continue to fly in the swarm.

It is particularly interesting that according to Kessel the formation of dancing swarms and the transfer of the initiative for copulation to the female "is a feature common to all the remaining stages in the series of mating patterns to be described."

Stage 4. In the fourth stage the male bears a wedding gift which, as before, consists of a prey, but now something has been added. Pasted haphazardly here and there against the prey are shiny viscid globules or, as in certain cases, the prey is more or less entangled in silk threads. . . . In this stage it is still the prey which serves as the mating stimulus for the female, the simple wrappings being unimportant in this regard. These entangling structures help to quiet the newly captured prey and doubtless that is their sole function.

Stage 5. In the fifth stage the simple wrappings of the fourth stage have been elaborated to form the complex structure which we call a balloon, and this structure shares with the relatively large prey the function of stimulating mating.

Stage 6. In the sixth stage the prey again serves as the stimulus for the male to construct the complex balloon. Now, however, the prey no longer serves as a food for the female. It is somewhat smaller than in the preceding stage and it is probable that the male consumes what fluid it contains when he captures it. At any rate it is no longer edible when the sexes embrace and the male passes the balloon to the female. In this stage, however, the prey, still being conspicuous enough to be easily noticed, doubtless continues to share with the balloon the function of stimulating copulation.

Stage 7. In stage 7 the prey is yet smaller; in fact it is always minute and so delicate that it would seem to be useless as food even for the male when first captured. The abdomen is always collapsed and is often fragmented into tiny pieces which are plastered into the front surface of the balloon. One may conclude that in this stage the prey still serves as the stimulus for the male to construct the complex balloon inasmuch as he always begins by catching the prey. But certainly the prey no longer serves as food for the female; moreover, because of its minute size and the fact that the female pays no attention to it while in the process of manipulating the balloon, it would seem that it has entirely lost its significance in courtship. It may be assumed, therefore, that in this stage the balloon remains as the sole stimulus for copulation.

Stage 8. In this last stage of evolutionary sequence the male no longer needs to capture a prey in order to be stimulated to construct a balloon. So the gift package includes no prey whatever, not even a dried-up, inedible one such as is present in stages 6 and 7.

Kessel names species and species groups of Diptera representing each of the eight stages. This example shows not only how investigations of comparative behavior would have to proceed in order to be useful in phylogenetic systematics, but also the limitations of this method in practice. No doubt it is far more difficult and time-consuming to make a complete survey of only the mating activities in a large group of species than of the morphological characters. Moreover, with behavioral characters it would seem to be even more difficult than with morphological characters to distinguish true "synapomorphy" from convergence and parallelism, and to determine how far reversibility of evolution occurs. The benificiary from a closer connection between investigations of comparative behavior and phylogenetic systematics will probably for a long time be primarily comparative behavior. It can gain a deeper understanding of its objects from the relations between the different modes of behavior and the phylogenetic relationships of their bearers as determined by other methods.

To an increasing extent, the chromosomes are being used in attempts to disclose phylogenetic relationships between species and species groups. These attempts are particularly interesting because they take into account—at least partially—the viewpoint that is decisive for phylogenetic systematics: it is not the degree of similarity in chromosome structure, but rather synapomorphy, that is decisive for assuming phylogenetic kinship. Most important is the method that infers phylogenetic kinship from the occurrence of "crossover inversions." "Suppose we observe in different strains the arrangements ABCDEFGHI, AEDCBFGHI, and AEHGFBCDI. The first can arise from the second or give rise to the second through a single inversion. The same is true for the second and the third. But the third can arise from the first, or vice versa, only through the second arrangement as the probable intermediate step in the line of descent. If we find in natural populations of some species only the first and the third arrangements, it is probable that the second remains to be discovered, or at least that it existed in the past. If all three are actually observed, the probability of the first and the third being related through the second becomes almost a certainty. To recapitulate, the phylogenetic relationship of the three gene arrangements indicated above is 1 2 3, or 3 2 1, but not 1 3. With independent and included inversions, no determination of the sequence of origin is possible; they are phylogenetically ambiguous. The existence of previously unknown gene arrangements in *Drosophila pseudoobscura* and *D. azteca* was predicted with the aid of the theory of overlapping inversions, and most of these predictions were subsequently verified by discovery of the requisite inversions in nature (Dobzhansky and Sturtevant 1938, Dobzhansky 1941)." (Dobzhansky 1951.)

Unfortunately this method is applicable only to the Diptera, and only to a limited extent. The viewpoint that only synapomorphous agreement indicates that several species represent a monophyletic group is also expressed in other studies of dipteran chromosomes. White (1949), for example, bases close relationship on the loss of chiasmata in males: "the possibility that loss of chiasmata in the male originated at several distinct occasions in the phylogeny of the Diptera (i.e., that the character is a polyphyletic one) seems to us an extremely remote one." For this particular character, however, it must be strongly doubted that White's assumption is correct. The occurrence of this character in several families of Diptera in several cases so obviously contradicts the phylogenetic relationships based on other criteria that we must assume that it did in fact originate several times by convergence. But this only shows that the use of chromosomes in phylogenetic systematics has the same limitations as the use of all other holomorphological characters and the transformation series that can be erected for them.

The validity of the criteria for determining the direction in which such transformation series are to be read is limited by several phenomena that will now be discussed.

(3) *Reversibility of evolution.*—The first prerequisite for unrestricted validity of the criteria discussed above would be strict irreversibility of evolution (transformation). The assumption of a more or less strict irreversibility of evolution is usually called "Dollo's law." Tillyard (1919) discussed it in detail under the name "Meyrick's law," and Sachtleben (1951) pointed out that Meyrick (1884) presented the hypothesis of irreversibility even prior to Dollo.

The fact that almost any organ that once appeared in a transformation series as a new apomorphous character may later be reduced to the point of complete disappearance shows that retrograde evolution can take place. However, the possibility that characters that have disappeared may reappear again is probably often underestimated.

Remane (1952) made a detailed study of the law of irreversibility. He came to the following conclusions:

1. Organs may be reduced or eliminated at any time.

2. Simple and individual form characters reappear in the course of phylogeny. It is therefore inadmissible to apply the law to proportional relationships, quantitative differences, etc.

3. Form characters may reappear at any time during phylogeny in cases where evolution resulted from genetic mutations. The "backward steps" then correspond to the back mutations.

4. Several and moderately complex form characters may reappear in the sphere of ecological types and analogies.

5. If only single stages (e.g., "adult stages") are considered, even complex structures may reappear if they were present without interruption in the juvenile stages. There may be cases in which a character reappears even when it was no longer visibly, but only potentially, present in ontogeny. In the same way, structures of an individual organ in a meristic system (teeth, vertebrae, leaves, etc.) may reappear if these structures are still present in other partial organs of the system.

6. No case of the reappearance of complex homologous organs, or of the total organization of a species, has been demonstrated. Therefore Dollo's law applies without restriction in this sphere.

Point (6) suggests a way of dealing with the difficulties that arise in systematics when a certain character, if traced back through a transformation series, appears in several species without these species having taken over the character in unaltered form from an ancestor that is common only to them. The phenomena of convergence and parallelism may be dealt with in the same way.

(4) *Convergence.*—In the systematic literature it is customary to speak of "convergence" when the same characters occur in different species but apparently were not taken over from a common stem species. The frequency of such convergences is estimated very differently: "Practical experience shows that convergence is relatively rare" (Maslin 1952), whereas Tuxen (1958) says "Convergence, I fear, is a much more common feature than is generally supposed." Such differences of opinion are very probably based on different concepts of convergence. Thus according to Remane (1952) it is characteristic of the "phylogenetic epoch" to regard analogy and convergence as synonymous. Remane says further that "organs of corresponding or similar structure that are not homologous are called analogous." Under such definitions analogies (and hence convergences too) would obviously play no very great role in the practice of systematics. Analogous characters whose agreement is based on convergence may then relate to very different transformation series for which there is no common initial condition realized in a single stem species. For the exclusion of such convergences in the disclosing of phylogenetic relationships, see what we said above (p. 93) on homology and the criteria of homologies. The assertion that convergences may appear in great numbers thus refers not to convergences in the true sense, but to what are often called "parallelisms" and "homoiologies."

(5) *Parallelism.*—"Although parallelism is radically different from the concept of convergence, the two are so frequently confused in common usage that the contrast may be emphasized. In convergence, two forms with similarities in directly adaptive structures (as wolf and marsupial 'wolf') have come from radically different ancestors with basically different patterns of organization; in true parallelism, both habitus and heritage are similar; the ancestral types were closely related, and evolutionary progress, stage by stage, has been closely comparable in the two or more lines concerned" (Romer 1949; similarly Heberer 1953). To parallelism in the broader sense belong above all those cases in which characters certainly absent in the stem species of a monophyletic group occur independently in the successor species.

Plate (1928) uses the term "homoiologies" (in contrast to true homologies in the narrow sense) for corresponding characters that occur in narrow kinship groups but which nevertheless developed independently in their bearers. True homologies, as is well known, are character correspondences that were actually taken over from the common ancestors as such, even though with partial alteration that still permits recognition of the common ground plan. "Homoiology describes a form of homology in which the particular character has been acquired independently by close relatives. It corresponds to the homoiogenesis of Eimer."

The eyestalks of the Diptera offer a very characteristic example of homoiology. This peculiar character occurs in a whole series of Diptera, in the families Platystomidae (genera *Achias* and *Laglaisia*, with about 25 species among the 1,200 stalkless species in the entire family), Trypetidae (genus *Pelmatops*, with one species among about 1,500 in the entire family), Richardiidae (two species of

Richardia and one of *Megalothoraca* among 150 species in the entire family), Pterocallidae (one species among 150 in the entire family), Tylidae (one species of *Anaropsis* among 150 in the entire family), and Diopsidae (all 150 species in the family). Thus the dipteran species or species groups characterized by eye-stalks are less closely related among themselves than to other species that have no eyestalks. They are nevertheless very similar to one another, and furthermore all belong to a broader relationship group, namely the acalyptrate Cyclorrhaphae. On the other hand, no forms with stalked eyes have been found in the many other groups of living Diptera. It can be no accident that so conspicuous and peculiar a character as stalked eyes occur only in a relatively small group of about 5,000 species (in which it apparently arose repeatedly) among the approximately 65,000 species of living Diptera.

Homoiologies occur not only among strictly morphological characters. They can embrace all conceivable form peculiarities, including physiological and psychological characters. An example is larvipary and pupipary in the higher Diptera. Larvipary, the development of embryos within the maternal body, has been observed in the most different species of the Calyptrata. Its occurrence in various small groups of calyptrates shows that it apparently arose repeatedly and independently. Larvipary has never been observed in the second large group of cyclorrhapids, the Acalyptrata. The same is true of pupipary, which must be considered a further development of larvipary in which the larva is retained even longer in the maternal body where it is nourished and finally born ready for pupation. Pupipary also occurs only in the Cyclorrhapha: in the pupiparous forms, which form a polyphyletic group that must be divided into two groups (Nycteribiidae-Streblidae and Hippoboscidae) that arose independently from muscid-like ancestors, and in the genus *Glossina*, which likewise probably arose independently from muscid-like ancestors. Thus here too we have a "character" that evolved independently at least three times in one major group.

Among homoiologies in the broad sense we may also include the occurrence of several characters (which may also appear independently) several times in the same combinations in a particular group of organisms, although we cannot assume that these identical combinations were taken over from common ancestors. Lindner (1939) has pointed to such a phenomenon in the Stratiomyidae. Here "the combination of steel-blue ground color of the body with yolk-yellow of the head" has appeared independently in various subfamilies. According to Lindner it occurs in the entire genus *Cyphomyia* among the Stratiomyiinae, in the entire genus *Patagiomyia* among the Hermetiinae, and in the species *Hypelophrum cyphomyioides* among the Pachygastrinae.

Homoiology occupies a peculiar intermediate position between true homology and convergence. It is also of great importance in the problems surrounding mimicry and in many other important biological questions. Many attempts have been made to explain it. Kiriakoff (1956) pointed to Vavilov's "law of the homologous series in heredity." Many cases may be explainable by the laws of allometric growth. In the titanotheres phylogenetic increase in body size is accompanied by "increase in the size of the horns, not only absolutely but also relative to body size. According to the principle of allometric growth, below a certain body size the allometrically growing organ (in our case the horns) would not appear at all. This is true of the earliest and smallest titanotheres. Although these were too small to form horns themselves, they were able to transmit the hereditary factors for horn formation to the groups that descended from them, which consequently developed horns independently of one another as soon as the body reached the necessary size" (von Bertalanffy, II).

One speaks of parallelism in the narrow true sense if different transformation series that arose from a common primary condition at first diverge, but later progress in parallel or finally even convergently. This can give the impression that several such phylogenetic series tend independently toward the same or a similar terminal condition without any recognizable external cause. An almost endless number of examples can be cited from all groups of organisms. A few random examples may be mentioned. According to Rüschkamp (1927) there is in the Chrysomelidae a general tendency for all "orinoid" forms to develop into "timarchoid" forms. According to James (1936) there are general tendencies toward specialization in the stratiomyiids: reduction and modification of antennal joints, concentration of veins along the costal margin of the wings, weakening and reduction of the veins at the posterior margin of the wings, a tendency toward shortening and broadening of the abdomen as well as reduction of segments. Schindewolf (1942) says that "the whole enormously long phylogeny of the stone corals is dominated by a direction of evolution that controls everything else: replacement of the original bilaterality of the septal apparatus by a hexameric radiality." Fittkau (1960) showed that parallel development of a great number of characters in chironomids is explainable by the decrease in body size that has taken place in several closely related groups in this family, apparently because size reduction has physiological and ecological advantages.

Today the term "trend" is commonly used for long unidirectional transformation series that formerly were often called "orthogenetic series." Numerous studies, particularly in paleontology, devote special attention to them (cf. also Stammer 1958). They have attracted attention especially because the question of how they are to be explained in genetic terms seems to be of particular significance for the factor problem of the theory of evolution (see the symposium on trends in evolution, Zool. Anz., p. 162, 1958). The question of how parallelisms in the narrow or broad sense are to be explained genetically is not of such great importance to phylogenetic systematics, for which it is a question of finding criteria that make it possible to decide whether or not the occurrence of identical characters or whole character complexes in different species is based on the fact that these were taken over from one stem species that is common only to these species (synapomorphy).

It has been variously recommended that a choice be made among characters used in systematic work, disregarding those that are particularly subject to adaptive change because of their importance in the life of organisms and consequently are especially likely to form parallelisms and convergences. "If the diagnostically useful characters of the different category groups (species, genera, families, etc.) are compared it is evident that their functional values do not correspond at all to the level of the particular category. Quite the contrary." "The completely unimportant, which offer no point of attack to any transforming forces and on which heredity can act completely undisturbed, are what in broad measure offer constant characters and consequently are disproportionately important in systematics." (Michaelsen 1935). Diels (1921) calls such "functionally independent" characters "constitutive characters," and cites Darwin's statement: "It can be considered a general rule that the less an organ is designed for specific physiological purposes, the greater its importance in classification."

This contradicts other recommendations, demanding that systematics consider all characters, ignores the fact that by no means all parallelisms appear to be adaptively determined, and finally neglects the fact that it is not the characters themselves that are important in systematics, but their relations to other characters in a transformation series—whether they represent plesiomorphous or apomorphous conditions. All these difficulties resolve into the requirement that characters cannot be considered in isolation, even in regard to transformation series of other characters.

In deciding whether *different* characters of several kinds are to be regarded as homologous, and therefore generally comparable with one another for the purposes of phylogenetic systematics, it is a question of determining whether they can be regarded as transformation conditions of a character that was present in a different condition in a stem species, which did not have to be the stem species of only the compared species. But in deciding whether *corresponding* characters of several species are to be regarded as synapomorphies, convergences, homoiologies, or parallelisms we must determine whether the same character was already present in a stem species that is common only to the bearers of the identical characters.

The method by which this can be determined is the same in both cases. It depends ultimately on bringing the characters in question into relation with other characters in the same species. Whether corresponding characters are to be regarded as synapomorphies, convergences, homoiologies, or parallelisms is decided as follows. If it must be determined whether the presence of a corresponding character a in the species B and C rests on synapomorphy, naturally it is assumed that symplesiomorphy has already been ruled out. Thus we must be convinced that the condition in which the character is found in all other species is more strongly plesiomorphous. Then one investigates another character, b, in the same species. If it is found that character b is likewise present in species A and C in an apomorphous condition (compared with A and other species), then the probability that the character is also synapomorphous is also increased.

But it may also turn out that character b is indeed present in species C, but is also present in another species A in a corresponding condition that must be interpreted as apomorphous, whereas it is absent in species B, which in regard to this character appears to be relatively plesiomorphous compared with A and C. Thus in this case the indications from characters a and b are contradictory with respect to the phylogenetic relations of species A, B, and C. It then becomes necessary to recheck the interpretation of characters a and b with respect to the following major possibilities:

(1) It was erroneous to interpret one or both characters as "plesiomorphous or apomorphous."
(2) Both characters may in fact represent apomorphous conditions, but the assumption that the presence of the apomorphous conditions in two different species is based on synapomorphy for one of the two characters (a or b) is erroneous. Either the presence of character a in B and C, or the presence of b in A and C, is to be interpreted as homoiology or parallelism.
(3) Either character a in C is not at all homologous to character "a" in B, or the characters b and "b" in A and C are not homologous. The apparently corresponding characters were not comparable to begin with.

Concerning the conclusions that must be drawn by phylogenetic systematics, possibilities (2) and (3) do not differ from one another. This explains and justifies the general looseness in the distinction between convergence and parallelism (including homoiology); "both refer to similarities that were evolved independently" (Bigelow 1958).

The common occurrence of parallelisms and homoiologies, if not of pronounced convergences, indicates the necessity for phylogenetic systematics to take into account as many characters as possible in deciding kinship relations. The more certainly characters interpretable as apomorphous (not characters in general) are present in a number of different species, the better founded is the assumption that these species form a monophyletic group.

But it often happens that only one character can certainly or with reasonable probability be interpreted as apomorphous, while other characters are either obviously plesiomorphous or in the present state of our knowledge not certainly identifiable in the group in question. In such cases it is impossible to decide whether the common character is indeed synapomorphous or is to be interpreted as parallelism, homoiology, or even as convergence. I have therefore called it an "auxiliary principle" that the presence of apomorphous characters in different species "is always reason for suspecting kinship [i.e., that the species belong to a monophyletic group], and that their origin by convergence should not be assumed a priori" (Hennig 1953). This was based on the conviction that "phylogenetic systematics would lose all the ground on which it stands" if the presence of apomorphous characters in different species were considered first of all as convergences (or parallelisms), with proof to the contrary required in each case. Rather the burden of proof must be placed on the contention that "in individual cases the possession of common apomorphous characters may be based only on convergence (or parallelism)." We discussed above how this proof is to be adduced. Tuxen (1958) probably misunderstood my heuristic principles when he wrote: "On another point I cannot follow Hennig, namely when he states that as long as convergence is not proved we should reckon with relationships: convergence, I fear, is a much more common feature than generally supposed."

Thus the question of whether kinship relations based on a single character or a single presumed transformation series of characters correspond to the actual phylogenetic relationships of the species is tested by means of other series of characters: by trying to bring the relationships indicated by the several series of characters into congruence. In the final analysis this is again the method of "checking, correcting, and rechecking" (see above).

This method is particularly applicable to organisms that undergo ontogenetic metamorphosis. The various stages of metamorphosis may be treated as if they were independent organisms, a separate phylogenetic system erected for each, and then an attempt made to bring these systems into congruence by the method discussed above. Unfortunately even here the concepts of similarity or of morphological or typological relationship on the one hand, and of phylogenetic relationship on the other, are not distinguished with the necessary sharpness. It must be recognized in principle that the requirement of complete congruence between larval and imaginal systems is theoretically justifiable only in phylogenetic systematics. Phylogenetic kinship exists only between species and species groups, not between larval and imaginal stages of these species. Only the morphological result of evolution differs under certain circumstances between larvae and imagines. The degree of morphological difference or similarity among several species may indeed differ greatly in the various stages of metamorphosis. From this we can conclude that it is impossible to bring into complete congruence larval and imaginal systems that attempt to express the degree of morphological similarity (typological or form relations) of these different stages of metamorphosis. Since phylogenetic systematics evaluates morphological differences and correspondences in a different way (not according to their magnitude), with proper evaluation it must always be possible to bring larval and imaginal systems into congruence.

Also erroneous is the idea that the larvae (of insects) are less important for systematics than the imagines because, on account of the caenogenetic or imaginifugal reconstruction of many characters (in which early ontogenetic stages become larval), they do not sufficiently recapitulate the phylogenetic development of these characters. "The degree to which their ontogeny recapitulates phylogeny, however, is precisely the scale by which the taxonomic significance of their characters is measured, so that their importance for taxonomy depends largely on the measure in which they are phyletic. As the typical larva thus does not represent an earlier phylogenetic stage of the winged insect, the larval characters cannot be regarded as a general rule to be of overriding importance for taxonomy" (van Emden 1957). These considerations relate, however, only to the importance of larvae in determining the direction in which particular transformation series of corresponding imaginal characters must be read (see p. 95). The special importance of larvae in phylogenetic systematics lies in a completely different area: larvae and imagines are the bearers of different series of transformation characters, and these must be brought into congruence in order to determine the phylogenetic relationships of the species to which they belong. "Rapid evolution of one stage, without comparable phenotypic change in the other, or with phenotypic changes having no obvious selective value in the other, is not fundamentally different from another common phenomenon, namely rapid evolution of one structure or group of structures while the others remain little changed. In this light it is easy to see that larval characters, and bionomic characters as well, are merely more characters, of no more and of no less systematic value than conventionally used characters" (Michener 1953).

In many cases the asserted incongruence between larval and imaginal systems is based on the fact that only the degree of similarity is considered, without raising the question of whether the similarity rests on symplesiomorphy or synapomorphy (Figs. 33-36). For phylogenetic systematics this is not an actual incongruence, but merely the result of a method that from its standpoint is false. "Morphologically the larva of *Micropteryx* is more closely related to the Mecoptera (= *Panorpata*) than to the moths, in which the adult has been placed, and is in fact less closely related to these than are the caddisflies, so that *Micropteryx* must be separated from the panorpoid stem before Trichoptera and Lepidoptera became differentiated" (van Emden 1957). Here the logical error of the metabasis is very clear: morphological "relationship" (of the larvae) is treated in the conclusions as if it were phylogenetic relationship.

Nor should we speak of incongruences if certain monophyletic groups of species can be set up, according to our scheme of argumentation (Fig. 22), only on the basis of larval characters, others only on pupal characters, and still others on imaginal characters. The result is not incongruent larval, pupal, and imaginal groups, but only monophyletic groups that cannot be justified as such in one or another stage of metamorphosis because for this stage of metamorphosis there are no known characters that can with certainty be interpreted as apomorphous (Fig. 36).

Naturally all this is true not only for animal groups with ontogenetic metamorphosis. It also applies to groups with pronounced alternation of generations: in the Hydrozoa the classification of the medusae is still rather independent of that of the polyps. For entire families of medusae we still do not know to which polyp families they belong as alternation of generation forms. This group presents an opportunity to test the efficacy of the methods of phylogenetic systematics by using them in all strictness first to produce independent classifications of the medusa and polyp generations. Then the "incongruences" between the two classifications would have to be tested according to the viewpoints sketched above.

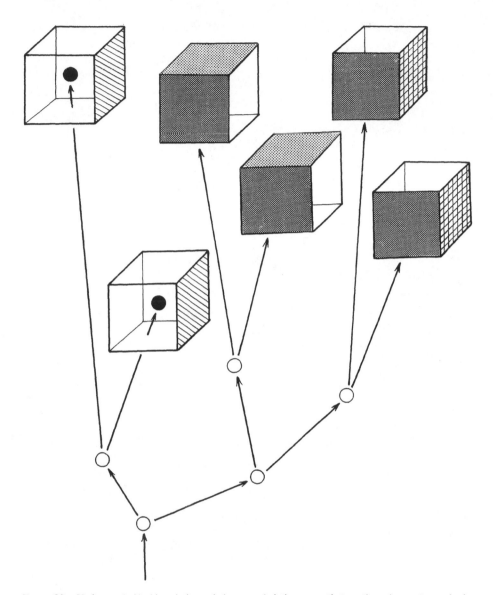

Figure 33. Phylogenetic kinship relations of six monophyletic groups that go through a metamorphosis. The imaginal characters are shown on the anterior faces of the cubes, larval characters on the right faces, and pupal characters on the upper faces. For all characters it is assumed that they occur in two transformation stages: plesiomorphous (white) and apomorphous (shaded). Compare with Figs. 34 and 35.

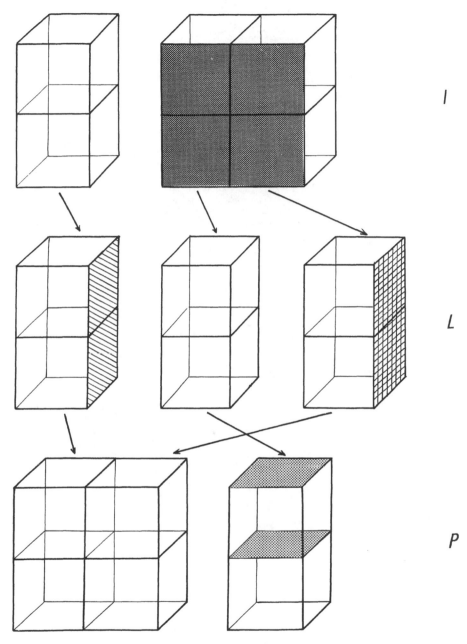

Figure 34. "Incongruence" between imaginal system (above, I), larval system (middle, L), and pupal system (below, P). Such "incongruence" results if the different metamorphic stages are classified according to morphological similarity, without considering whether the "similarity" is based on symplesiomorphy or synapomorphy.

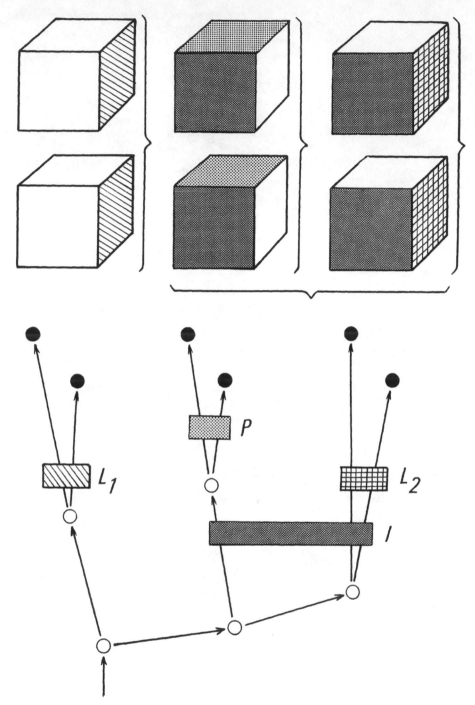

Figure 35. Grouping of the six species or morphological groups represented in Figs. 35 and 36 with the correct methods of phylogenetic systematics, i.e., the fact that monophyly can be based only on the demonstration of synapomorphous characters. Then it is conceivable that some monophyletic groups could be based only on characters of the larval stage (L_1 and L_2), others only on characters of the pupal stage (P), and still others on characters of the imaginal stage (I). The possibility of finding **diagnostic** characters for **all** stages of metamorphosis is thereby not at issue.

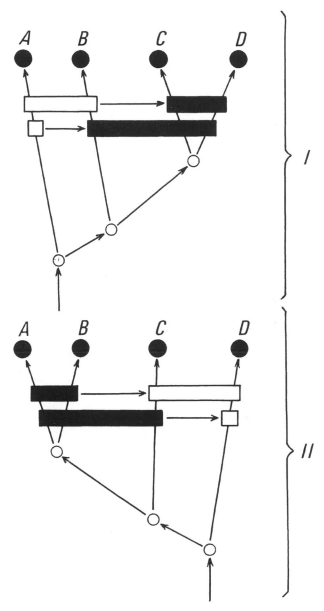

Figure 36. True "incongruence" between imaginal (I) and larval (II) systems: by demonstration of apparently synapomorphous characters different monophyletic groups are "established" in the imaginal stage than in the larval stage. In this case the stages of expression of the characters either in the imagos or in the larvae have been wrongly interpreted (the black bars indicate in the one or the other stage the plesiomorphous and not as assumed the apomorphous stage of expression), or the properly interpreted apomorphous stage of expression has been reached in the imaginal or larval stage by convergence.

With the questions of how it is possible to determine whether characters in different species belong to one and the same series of phylogenetic transformations, if so and we are dealing with corresponding characters, whether they are to be interpreted as plesiomorphous or apomorphous, and finally, if apomorphy can be made probable, the characters are to be interpreted as synapomorphies, parallelisms, or homoiologies; then we are already in the midst of the question concerning the truth content of the assumptions by which phylogenetic systematics undertakes to determine the phylogenetic relations among species.

Our discussion has shown that there is no simple and absolutely dependable criterion for deciding whether corresponding characters in different species are based in synapomorphy. Rather it is a very complex process of conclusions by which, in each individual case, "synapomorphy" is shown to be the most probable assumption. This fact appears to justify those who contend that systematic work has moved far from the area of true scientific research, which is characterized by the use of exact methods, into the domain of intuitive-emotional decisions. According to this example, in systematic work overriding importance is attached to the often-cited "systematic tact" or "morphological instinct" (Naef) and other subjective and unteachable methods.

"Already long before Darwin," according to Roux (1920), "a so-called tact was demanded of systematists. With its help they were supposed to determine almost instinctively the relationships among very heterogeneously constructed forms." Mez (1926) also reproaches modern "transcendental systematics," which he would replace with serum diagnostics, with "no longer working with clear morphological characters but with the systematic 'feeling' of the investigator. This it must do because morphology is completely incompetent to deal with questions of descent and lacks any objective way of proving its deductions." Statements of this kind have often lead to an emphatic skepticism of the results of systematic and phylogenetic work in general. "We may now develop a certain idea of the course of phylogeny in the salpids, but we are on guard against believing that the unascertainable phylogeny actually followed this course" (Ihle, in Kükenthal-Krumback 1935).

These and similar statements do not do justice to the capabilities of the methods of phylogenetic systematics. Almost all start from the idea that systematics must be content with deducing phylogenetic relationships from the degree of similarity between different species, and such a way of reaching conclusions is in fact very problematical (see p. 76). But as we have seen, this is not the true method of phylogenetic systematics. Phylogenetic systematics starts rather from the conviction that all correspondences and differences between species arose by an alteration of characters of common stem species in the course of phylogeny. It attempts by criteria not based on the degree of correspondence and difference, to determine the direction in which the changes took place. Thus the attempt to reconstruct the phylogeny, and thereby the phylogenetic relationships of species, from the present condition of individual characters and the presumed preconditions of these characters has the nature of an integration problem. In mathematics, the most exact science, according to Michaelis (1927), "integration . . . is an art . . . since one is often faced with the problem of combining, from the numerous possible manipulations, those that make possible the solution of the problem." This is precisely the situation in which the phylogenetic systematist finds himself. For solving his "integration problem" he also has available exact scientific methods, the discussion of which forms an essential part of the content of the present book. If, in determining which of these methods would make possible the solution of a particular problem, he is dependent on capabilities that do not lie in the realm of the learnable, this should not be objected to any more than when the mathematician finds himself in the same situation. According to Spann (1935), the mathematician Gauss once said: "I have the result, but I don't know yet how I got it."

The question now arises as to what guarantee the systematist has that the solution of his problem is really correct. Most previous theoreticians of systematic work evidently believed that in this respect phylogenetic taxonomy is in a basically different and less favorable position than other biological disciplines, particularly the so-called exact natural sciences, because the relationships among organisms that it attempts to determine are not actually existing causal relationships (in this sense "real"), but are based on past events. This notion rests on a basic misunderstanding of the criteria of "proof" and truth."

"That an 'assumption' is true is shown by its 'confirmation' within the thought relationship, inasmuch as it not only does not contradict this but fits into it harmoniously and remains constantly attached to the character of the demand" (Eisler, *Handwörterbuch der Philosophie*). In this sense, according to Müller-Freienfels (cited by Eisler), "true" is synonymous with "according to relationship."

But according to Gehlen (1940) not only the fact that an assumption can be fitted without contradiction into a system of relationships already known and considered valid, but also its "fertility" in disclosing new relationships must be regarded as a "criterion of truth":

in such a problem thinking acts by feeling out the questionable situation or object from different sides in order to determine what is known or can be added to what is already known. In the empty places so exposed, concepts are tentatively added that "break down" or reflect the subject, until a promising proposition is found. This is good or promising if it not only brings the unknown into a relation with the known, but particularly if it permits new possibilities to develop from the problem, which are then pursued in order to determine whether they can be confirmed. In the former case the whole process so far followed is legitimized backward and the system of insights thus developed is called "true" and now forms an invariant that can enter into the solution of new problems.

For phylogenetic systematics this means nothing more than that it too has at its disposal criteria with which the truth or correctness of its results can be tested, and that these criteria of truth are basically the same as those on which all other sciences must rely. Thus for phylogenetic systematics too they consist of testing whether the results of its work are "interconnected" and "fertile," i.e., whether they fit into the fabric of already recognized conformities to law and whether new conformities to law can be derived from them. This question can be decided primarily by determining whether the differently determined views concerning the phylogenetic relationships of particular groups of organisms are in agreement. This method was discussed above (p. 84). A very simple example may show even more clearly how it works. Suppose a geographer has obtained fragments of a topographic map of an unknown land. He will make every effort to reconstruct the map from the fragments. How can he succeed if the original map is unknown to him? He was not present when the map was torn up. The geographer must try to assign each fragment to its original place in the totality of all the recovered fragments. He will proceed by trying to find, for a portion of a river present on one fragment of the map, the adjoining piece of the same river on another fragment. If he directs his attention to a single geographic element in his map fragments, such as rivers, he is likely to go wrong. Thus the three sections of a river, a, a', and a" (Fig. 37) could seem to be upper, middle, and lower parts of this river, or he may interpret parts of different rivers as parts of the same river.

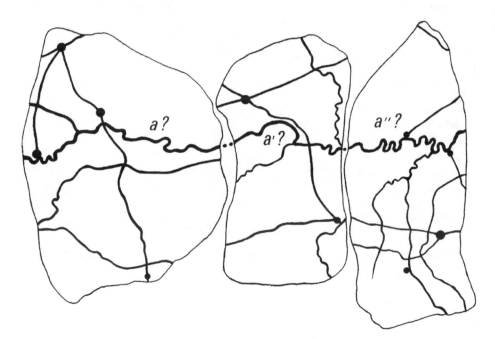

Figure 37. Criterion of veracity. In the reconstruction of a topographic map from several fragments, a, a', and a" could be interpreted as adjacent sections of a stream course. The remaining elements of the map, however, remain isolated. The joining of the map fragments is wrong.

His error becomes obvious if he considers other elements ("characters") of his map fragments. They remain isolated; pieces of roads and railroad lines do not join up (Fig. 37). But if all geographic elements are satisfactorily fitted together (Fig. 38) the geographer will be convinced that the fragments have been assembled "correctly," even though he did not know the original condition of the map.

The analogy to the methodology of phylogenetic systematics is clear. Perhaps this example will also clarify the difference between the methods of typological and phylogenetic systematics for those authors who have been unable to see it. Myers (1960), for example, writes: "I doubt exceedingly that there is today a really definable difference between phylogenetic and other kinds of taxonomy, short of pigeonholing by a nonevolutionist." Typological systematics rate differences and conformities as "essential" and "nonessential" or "less essential." But the yardstick by which they are rated may differ greatly. Cuvier, for example, according to Cain (1950), asked "which characters in animals are the most influential?" and made those he considered most influential the basis of his classification. Lamarck, de Candolle, and Caesalpino proceeded likewise, according to Cain. Modern authors often consider as most essential the magnitude of the mor-

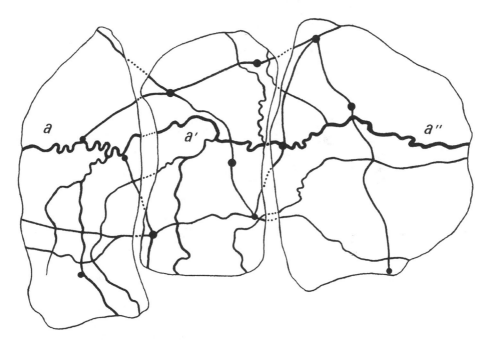

Figure 38. Criterion of veracity. If in the reconstruction of a topographic map from several fragments, a, a', and a'' are interpreted as adjacent sections of a stream, the other elements of the map also join to form a sensible illustration. The joining of the map fragments is correct.

phological differences (which they usually try to formulate in some kind of mathematical terms), and because they see in this a "measure of the amount of evolution" (Michener 1957) or the "results of evolution" (Bigelow 1958), they often believe—to judge from their terminology—that they are engaged in truly "phylogenetic systematics." But who could contradict an author who does not regard morphological differentiation as the essential result of evolution: with Stammer (1960) when he says "for a zoologist, it seems to me, the organizational type of an animal is the essential feature," and consequently supports Huxley's proposal to oppose man (as the Psychozoa) to all other organisms because of the psychic result of human evolution? Who could contradict an author to whom the difference between the Pterygota and its sister group among the "Apterygota" (see Fig. 69) seems so "considerable" (regardless of how great this difference may appear in mathematical terms) that he regards it as one of the most essential results in the evolution of the Arthropoda and consequently would not include the "Apterygota" and "Pterygota" in a class Insecta (Boettger 1958).

In the final analysis the question of what is an "essential" result of evolution cannot be argued objectively. But even from the standpoint of the methodological question, all attempts to make the "result" of evolution (whatever this may mean) the basis of systematic classification stand in sharp contrast to phylogenetic systematics. Consequently we label all such systems "typological," despite the possibility of making even finer typological distinctions between different kinds of typological systems. Phylogenetic systematics differs from all of them even in method, since in phylogenetic systematics the question of which agreements and differences—whether quantitative or qualitative—are to be regarded as essential expressions of the results of evolution is indifferent. Only the position of the characters within transformation series is determinative, and phylogenetic systematics tries to fit in all available characters with the aid of the methods discussed above. In so doing, it determines the probability with which the bearers of conforming or different characters can be regarded as members of monophyletic groups. To return to our geographic example, obviously the probability of reconstructing the map correctly increases with the number of usable elements (pieces of rivers, roads, railroads). For phylogenetic systematics this means that the reliability of its results increases with the number of individual characters that can be fitted into transformation series.

It is evident that the probability of the veracity of an assumption concerning the phylogenetic relationships of systematic groups increases with the independence of the relations between the organisms among which conformities to law can be recognized.

But all configurational relations are interconnected. Therefore in order to test the correctness of the phylogenetic relationships deduced from them we must above all determine whether the relationships among organisms that are indicated by nonmorphological data agree with the deductions from the morphological data. The distributional relations of the organisms can be used for such testing—they were likewise brought about by phylogeny, but are only remotely connected with the form of the organisms. The reliability of phylogenetic relationships derived from certain structural relations can therefore be regarded as especially reliable if the distributional relations of the same organisms also show an orderly structure.

Naturally we can proceed equally well in the opposite direction and make the distributional relations themselves the starting point for deriving phylogenetic kinship relations. The reliability of the results can then be tested against the resulting holomorphological consequences. Thus the analysis of the distributional relations of organisms—the "chorological method"—is an important aid in taxonomic work.

The Chorological Method. In discussing taxonomy it was shown that there are close relations between the species and space. Every species originally occupies a certain area, and the breaking up of a species into several reproductive communities usually, if not always, is closely related to the dispersal of the species in space. Consequently the distribution of the closest reproductive communities in space could also be used as a criterion for the phylogenetic relationships between them (the "vicariance criterion"). It was recognized long ago that the spatial relationships between organisms can be used for determining phylogenetic relationships—and thus for solving the problems of phylogenetic systematics—in cases where the morphological method does not suffice.

In botany von Wettstein seems to have been the first to point explicitly to the importance of the geographical method for systematics. He used it in working over certain groups (e.g., *Euphrasia*). The use of the geographical method in zoology has become known particularly through the books of Rensch. It is almost exclusively restricted, however, to the lower taxa. Especially in the taxonomy of birds and mammals, the opinion is now rather general that vicarying reproductive communities are to be regarded as subspecies of a single species. The term *rassenkreislehre* is used for this method and all questions surrounding its principles of application and results. This fact alone indicates that geographic vicariance and the possibility of using it in phylogenetic systematics are regarded as a peculiarity limited to relationships among the lower taxa. This is not true at all. There are also relationships between distribution in space and the systematic classification of the higher taxa.

For the relationships between the chorological distribution and phylogenetic relationships in the higher taxa it may be taken as a ground rule that species groups belonging to a community of descent are restricted to unit areas that are to a certain extent unbroken. In the chorological method it plays about the same role as the statement that the more similar two groups are the more closely they are related phylogenetically does in morphological methods.

The example of the genus *Myennis* shows how this rule can be used in determining phylogenetic relationships. The dipteran groups known under the names Otitidae and Pterocallidae each include about 150 known species. These two groups are undoubtedly very closely related, and are not always easily distinguish-

able from one another. They have distinctly different ranges, however: the Otitidae are almost exclusively Holarctic, the Pterocallidae almost exclusively Neotropical. Until recently it appeared that the pterocallids were represented in the
Palaearctic by the genus *Myennis* (5 species). Neotropical and Palaearctic species were once united in this genus, but later study showed that the Palaearctic
species form a separate group not closely related to the Neotropical species
(which today comprise the genus *Neomyennis*). The strange occurrence of a
pterocallid genus (*Myennis*) in the Palaearctic still remained. The presence of
Myennis far outside the range of all the other pterocallids led to a re-examination
of the genus, particularly since the great morphological similarity between the
pterocallids and otitids led to the suspicion that a false evaluation of the similarity relationships was responsible for the inclusion of the genus in the Pterocallidae.
In fact it turned out that the species of the genus *Myennis* differ from all other
pterocallids and agree with the Otitidae in a previously unobserved character,
the armament of the male genitalia. On the basis of this, *Myennis* was transferred
from the Pterocallidae to the Otitidae, in whose range it occurs.

The interpretation of Kiriakoff (1956) of the so-called subfamily Brephidiinae
of the lepidopteran family Lycaenidae rests on the same considerations. Kiriakoff
starts from the fact that America and Africa do not form a single area of distribution, and that "recent investigations have shown almost irrefutably that no
land connection between South America and Africa across the south Atlantic existed, at least during the Middle Mesozoic or later." He therefore considers it
improbable that the subfamily Brephidiinae is a monophyletic group. From
Kiriakoff's statements it seems that the correspondence between the American
and African species rests on symplesiomorphy. If this is true it could not be attributed to the action of Vavilov's law, as Kiriakoff assumes.

There is no need to give further examples. Any systematist revising a supposedly monophyletic group would check the affiliations of a species group that
occurs far outside the otherwise continuous range of the main group. He would
test particularly critically whether it really belonged in the group in question.
This he would do by investigating, by the criteria discussed above, whether the
morphological characters that had determined its systematic assignment must
actually be considered synapomorphous correspondences.

Within the continuous range of a monophyletic group it is also possible, as
noted above, under certain circumstances with certain transformation series of
characters, to determine the direction in which it must be read.

We have already pointed out several times that speciation apparently always
goes parallel with a progression in space. Perhaps this is expressed most conspicuously in the fact that the speciation process obviously goes through a stage
in which the daughter species are vicariant. This need not be vicariance in geographic space.

We can visualize two possibilities with respect to the relationship of the ranges
of the daughter species (or races at first) to the original range of the parent

species. In dichotomous speciation one of the two daughter species remains in the old range of the species, while the other inhabits the new area that has been acquired. The second possibility is that neither of the two daughter species remains in the original range because the group was forced by some kind of external circumstances to give it up. The question now arises as to what relationships there are between the "progression in space," which according to the above is connected with speciation, and the progression of the morphological characteristics of the daughter species.

We would expect a study of the geographic rassenkreise to give the most reliable answer to this question. For the purposes of the problem dealt with here we may distinguish between rassenkreisen in which the vicarying subspecies are arranged in approximately linear succession—in which extension evidently took place only in one main direction—and rassenkreisen in which extension took place in all directions from a center of distribution. In the first case (the so-called chains of races) we often find that one or several characters in each successive race are a further development of these characters in the preceding race, and that the direction of further development of the characters is the same. Consequently these are orthogenetic series, in which the direction of differentiation

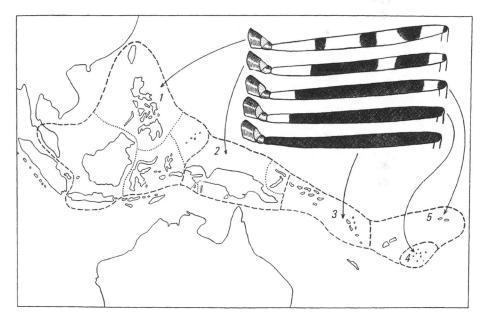

Figure 39. Pattern of the thigh in different subspecies of **Mimegralla albimana** (Diptera, Tylidae).

From top to bottom: 1, western group of races (**albimana, sepsoides, galbula, palauensis**); 2, New Guinea, Bismark archipelago, Key Islands (**contraria, keiensis, striatofasciata**); 3, Samoa (**samoana**); 4, Tonga Islands (**tongana**); 5, New Hebrides (**extrema**).

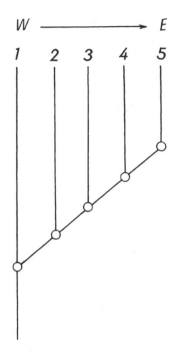

Figure 40. Diagram of the phylogenetic relationships between the five subspecies of **Mimegralla albimana** shown in Fig. 39. At the same time the stage series of increasing apomorphy in blackening of the thigh corresponds to the direction of spreading from west to east.

can scarcely be correlated with an approximately similar unidirectional change in living conditions. Examples of such unidirectional deviation are shown in Figs. 39, 40, and 41 for the dipteran genus *Mimegralla* and the reptile genus *Draco*.

Mimegralla albimana (Tylidae) breaks up into several subspecies in the area between the mainland of southeast Asia and the New Hebrides and Tonga Islands. These subspecies differ, among other things, in their leg markings, each subspecies to the east having darker leg markings than its neighbor to the west. This is clearly shown in Figs. 39 and 40.

In the flying lizards (*Draco*) there are two species (rassenkreise) distributed from the mainland of southeast Asia to the eastern limits of the Oriental region. In both rassenkreisen the most derived pattern on the flight membrane is found in the races farthest removed from the center of distribution of the rassenkreis (Fig. 41). In the rassenkreis *lineatus* the western races (*lineatus* from the Greater Sunda Islands, *beccarii* from southern Celebes) are characterized by primitive, relatively complete patterns on the flight membranes. The race occurring in north-

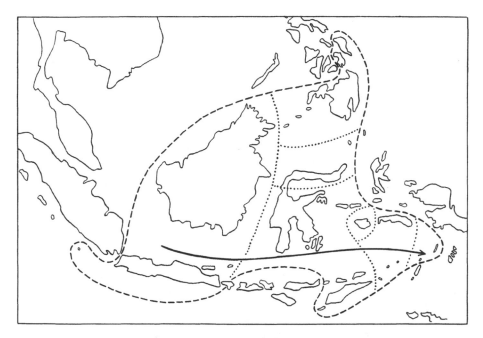

Figure 41. Range of distribution and supposed direction of spread of **Draco lineatus.** The general direction of spread corresponds to a stage series of increasing apomorphy in the pattern of the flight membrane (see text).

ern Celebes (*spilonotus*) has only vestiges of the pattern at the edge of the flight membrane closest to the body. In the most eastern race, *ochropterus* from the Key Islands, the flight membrane pattern is completely absent. The situation is even more striking in the other rassenkreis (*volans*). In the most easterly races (*boschmai* from the Lesser Sunda Islands, *reticulatus* from the Philippines) the partly very distinctly banded pattern of the western nominate form is fused into a reticulate pattern. It is very interesting that the forms from the Philippines and the Lesser Sunda Islands were until recently considered identical and were included under the same name. The two races apparently originated independently from the more western stem form, however, and so have nothing to do with each other in the immediate phylogenetic sense. Their nearly identical reticulate pattern is rather to be regarded as convergence (Hennig 1936a).

In all these examples the direction in which the species extended their ranges is known with adequate certainty: in all it was in general from west to east. This makes it possible to determine exactly the phylogenetic relationships between the individual subspecies. It must be assumed that each subspecies stands in a kinship relation of the first degree to all subspecies lying east of it, so that a hierarchy of degrees of phylogenetic relationship within the species, as shown in Fig. 40, is to be assumed. If the morphological differences are compared with this diagram of the degrees of phylogenetic relationship, it is evident that in each case the more easterly form shows a more advanced (more apomorphic) stage of development of the characters distinguishing them than does the neighboring form to the west.

But the concept of space in the systematics of the higher taxa need not be limited to geographic space any more than was the case in the differentiation of species (see p. 47). Rather, according to the first ground rule of the chorological method, we must demand that certain unbroken areas in "living space" be assumed for the higher taxa too. This in fact is usually the case. Everyone knows that most higher taxa, insofar as they are true phylogenetic units, have certain "ecological characters" in addition to the structural characters—just as species and subspecies do. In other words, they too occupy certain uniform areas of the living space.

This rule can also be used in the opposite direction in systematics. For example, we view with skepticism the suggestion—made most recently by Imms—that the Braulidae (bee lice) are closely related to the Chamaemyiidae; because all chamaemyiids are shield-louse parasites and therefore inhabit a peculiar and sharply characterized ecological niche in which *Braula* does not occur.

The occurrence of the higher taxa in certain ecological niches or zones is no doubt based, exactly as in species, on certain physiological, ethological, and in the narrow sense even morphological, peculiarities in their form. Consequently we may say that even in the higher taxa very definite dimensions of form (holomorphy) correspond to the different dimensions of the environment. Only the two or three dimensions of the environment that we assign to geographic space form an exception; no particular dimensions in the form of organisms corresponds to these. Although the division between ecological and geographic space is to a certain degree artificial and the concept of purely geographic space merely an abstraction, it must nevertheless be maintained—for reasons that have been discussed repeatedly—that the geographic dimensions of the environment are of particular importance to phylogenetic systematics. Fortunately in practice they can usually be separated sufficiently sharply from the other dimensions of the living space. Consequently in the following we will restrict ourselves essentially to them. We will never forget, however, that everything we say about the two or three geographic dimensions of the total environment also applies in the last analysis to the other dimensions, and thus to the total environment itself.

The rule that phylogenetically homogeneous groups of higher rank also inhabit fundamentally continuous areas should not be misunderstood to mean that such areas (a continent, for example) must be unbroken in the trivial sense. They may extend over broadly separated parts of different continents, and such cases are called "disjunctions." Darlington (1957), among others, has described the most important of them (Fig. 42). These are of particular importance in systematics because in a monophyletic group with disjunct distribution the partial areas are usually occupied by different partial groups. Species groups that arose from the stem species of a monophyletic group by one and the same splitting process may be called "sister groups." We can then show that there is often a vicariance relationship between monophyletic groups that represent such sister groups. We may find, for example, that the sister group of a Neotropical group is Holarctic or Australian (Fig. 42).

Up to now we have said that the task of phylogenetic systematics is to construct a system that contains only monophyletic groups, at the same time presenting all recognized monophyletic groups. Instead we may now say that its task is

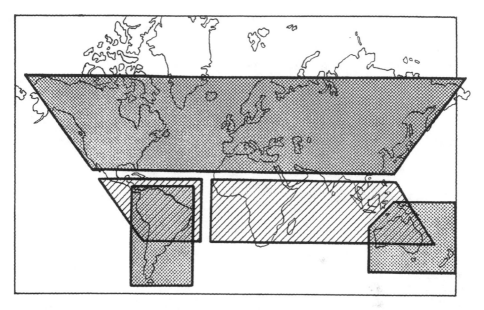

Figure 42. Some of the most important disjunctions (vicariance types of higher order).

to express all sister group relationships: every monophyletic group, together with its sister group (or groups), forms—and forms only with them—a monophyletic group of higher rank. Once a monophyletic group has been recognized, the next task of phylogenetic systematics is always to search for its sister group. The importance of the study of vicariance types is that it provides systematics with clues to the geographic region in which this sister group is to be found.

A little thought shows that sister groups must have the same absolute rank in a phyletic system. Consequently the study of vicariance types is also important in determining the absolute rank of systematic groups. This will be discussed further in the section devoted to this question.

The Paleontological Method. The recognition and systematic presentation of the phylogenetic kinship relationships between species and higher taxa is identical with determining the different time spans, regardless of how the latter are measured, that separate these species and higher taxa from their respective stem species. Paleontology deals with the life of past geological periods. We might assume, therefore, that it could provide systematics with direct information on phylogenetic ties and consequently be the most important accessory science in the taxonomy of higher taxa. Paleontologists have supported this idea even down to the most recent times. For example, according to Schindewolf (1942) "it should be self-evident that the actual course of phylogenetic differentiation, a process that lies in the past and therefore is a historic one, can be disclosed only by the historic records," records that are to be worked on by paleontology, according to Schindewolf.

The validity of this assumption is very limited, since the phylogenetic process as such is not supplied to the paleontologist. Animals of the past are not available to him as living organisms (i.e., in all characters of their holomorphy), but even in the most favorable cases only as more or less well-preserved carcasses or skeletons, very often merely as impressions of the outer body or even of body parts. Consequently paleontology may not be able to say much directly concerning the autogenetic relations between organisms from a particular geological horizon, or the phylogenetic relationships between these organisms and those living today. In grouping the known organisms (semaphoronts) of a limited time period into species (i.e., in recognizing the autogenetic relationships that existed between these organisms), the paleontologist is limited to the same methods as the taxonomist of recent species in cases where direct recognition of the autogenetic relationships is difficult, time-consuming, or for any other reason not feasible (see above, p. 66). In other words, he is limited primarily to comparative morphology and chorology. But the paleontologist is in a far more difficult position than the taxonomist of recent forms, since he is never able to study the entire structure (holomorphy) of the organism. A certain degree of reconstruction of the holomorphy is possible ("paleobiology"), but only on the basis of correlations that have been determined among the various structural characters in recent forms. These correlations are then transferred to the fossil forms, with a certain justification that will not be discussed here. Determination of species boundaries—in other words, reconstruction of the autogenetic relationships that may have existed among the fossil organisms—is possible only by transferring empirical knowledge gained from recent organisms. Often this is particularly true of the extent and character of individual variability.

It is immediately evident that the reliability of the results, in the reconstruction of both holomorphy and autogenetic relationships, will vary directly with the geological age of the organisms and the degree to which their structure resembles that of the recent forms from which the empirical knowledge was transferred. The older the fossils, and the less similar they are to the living forms, the less reliable will be a direct transfer of knowledge based on the living forms—attempts to determine species boundaries, for example, and to recognize possible metamorphisms, cyclomorphisms, and polymorphisms. Consequently it is not surprising that Beurlen (1937) notes that investigations on ammonites have in part led to complete resignation regarding the possibility of distinguishing species. We cannot interpret this, as Beurlen did, as militating against the general validity of the whole species concept. It is only an expression of the unsatisfactory position of paleontology with respect to the possibility of determining genetic relationships that existed between fossil semaphoronts.

The Rhachicerinae of the Baltic amber show how difficult it can be to clarify questions of species membership even in relatively recent fossils that are very similar to their living relatives. All 25 species of the single recent genus *Rhachicerus* show pronounced sexual dimorphism. From the Baltic amber six individual fossils (individuals, inclusions) belonging to this relationship group are known. Three are males and three females. The differences between the sexes make it certain that the sexual dimorphism observed in the recent species was also present in the fossil species. Further differences among the male fossils indicate that at least two fossil species must be distinguished. The genus *Lophyrophorus* has been erected for one of these, while the other has been placed in the recent genus *Rhachicerus*. Probably the three female specimens also belong to these two species; before the sexual dimorphism in the recent forms was known, they were united in a third genus, *Electra*. Even if they represent the opposite sex of the males of *Lophyrophorus glabellatus* and *Rhachicerus* sp. it cannot be determined with certainty which female belongs with which male. There is even the further possibility that the females—or at least one or another of them—represent a third species of which the males are still unknown. Consequently there is absolutely no way of deciding with certainty, among the small differences shown by the inclusions, what represents individual variation, what sexual dimorphism, and what is based on true species differences. Discovery of additional material, including copulating pairs, might make a decision possible.

Similar examples can be assembled from all animal groups. In the Foraminifera the presence of an alternation of generations, coupled with an external trimorphism, makes the species allocation of fossil forms in part extraordinarily difficult, according to Cushman (1933). Abel (1939) wrote: "Twenty years ago Franz Baron Nopsca expressed the suspicion that certain sex differences can be detected in the pelvis of dinosaurs, and that *Iguanodon mantelli* is to be considered the male and *Iguanodon bernissartensis* the female of one and the same species. Likewise on the basis of pelvic structure the genera *Saurolophus* and *Corythosaurus* are male and *Trachodon* female. This agrees perfectly with the interpretation of the differences in the skull as sex differences."

These facts and considerations lead to the conclusion, contrary to the overestimation of the value of paleontological methods and in agreement with Sewertzoff, Ehrenberg, Zimmermann, Naef, and others, that the paleontological method does not make possible direct determination of phylogenetic relationships, but promises results only in cooperation with other methods of phylogenetic systematics. "The paleontological record is a direct historical one, but it has the rather incidental drawback of superficial incompleteness, and the more profound disadvantage of the absence of a natural connection between the individual elements (fossils), which in general cannot be connected to stem or even ancestral series" (Naef 1919).

For the taxonomy of recent organisms, knowledge of the species boundaries of fossil forms would be of essential importance only if at the same time the phylogenetic connection between these fossil species and their recent descendants was obvious. It is apparent from the statements of the authors cited above that this is not usually the case. Zimmermann (1943) considers the effectiveness of paleontology in providing this information so unsatisfactory that he believes we will never be in a position to assume such a connection with certainty. "The ancestral series of the phylogenist are not only occasionally, but always, what Abel called 'step series.' In the characters that interest us the fossil forms represent the stage of development that had been reached at the time by those particular ancestors." In the area of what Zimmermann calls "character phylogeny" paleontology is in fact of great importance to phylogenetic systematics, even where phylogenetic systematics is limited to determining phylogenetic relations among recent species and species groups.

We have discussed in detail above that the determination of phylogenetic relationships by morphological methods rests on the assumption that it is possible to erect phylogenetic series of transformations of characters, and to distinguish plesiomorphous from apomorphous conditions within these series. To be able to do this successfully requires closely graded intermediate stages between the individual character conditions. In many cases these are available in the recent species of a group; many gaps can be filled by detailed ontogenetic studies. But often the gaps between the character conditions shown by the recent species are so great that it is not certain what arrangement would correspond to the actual phylogenetic development. In such cases the possibility of filling the gaps with the aid of fossil species may be of decisive importance to systematics.

A very simple example of this type is shown in Fig. 43. In the great majority of Bibionomorpha (Diptera) the radial sector of the wing is two-branched. In the basic plan of the dipteran wing venation it is four-branched (r_2, r_3, r_4, r_5). From the recent forms alone it is impossible to determine with certainty how the two-branched radial sector of the Bibionomorpha developed from the original four-branched radial sector. An answer to this question is one of the prerequisites for determining the relationship of the Bibionomorpha to other groups of Diptera. There are to be sure some very rare species of recent Bibionomorpha in which a third branch of the radial sector is present (Fig. 43E, *Pachyneura*), or in which a "transverse vein" can be interpreted as a vestige of such a third branch (Fig. 43B, *Cramptonomyia*). The mode of branching of these three veins does not

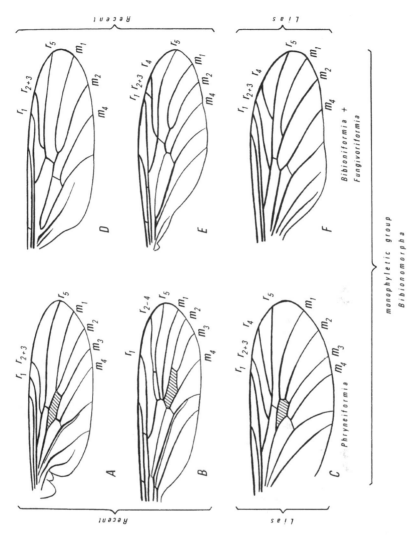

Figure 43. Paleontological method. Wing venation of A, **Phryne fuscipennis;** B, **Cramptonomyia spenceri;** C, **Protorhyphus stigmaticus;** D, **Amasia funebris;** E, **Pachyneura fasciata;** F, **Eoplecia primitiva.**

permit any certain interpretation of their homology, however. Such interpretation is possible with the aid of fossil forms from the Upper Lias (Fig. 43C, F). These fossil forms bridge the considerable morphological gap between the differentiation of the radial sector in the relatively plesiomorphous representatives of the recent Phryneiformia (Fig. 43A, B) and the Bibioniformia-Fungivoriformia

(Fig. 43D, E). Finally, they show that the close correspondence in the construction of the radial section between the relatively apomorphous representatives of the two groups (Fig. 43A, D) is based on convergence or parallelism, in any case not on synapomorphy. This proof emerges from the fact, evident in the Lias fossils, that in the Bibioniformia-Fungivoriformia a reduction of the discoidal cell preceded the reduction of the radial sector, whereas in the Phryneiformia the discoidal cell (Fig. 43, shaded) is retained even though the radial sector is reduced to two branches.

At the same time this example shows the importance of paleontology in deciding the question, discussed in detail on p. 96, of the sequence in which the transition from plesiomorphous to apomorphous conditions took place in the various transformation series. This question is often extremely important in systematics. Assume that we have a monophyletic group, A, that is well based as such on several apomorphous characters (a″, b″, c″, d″). Search for the sister group (p. 139) requires that we find another monophyletic group that agrees with group A in at least one apomorphous character. In such cases it often happens that our group A agrees with another group B in one apomorphous character (e.g., a″), and with a third group C in another (e.g., b″). According to our discussion above, the only possible assumption is that either the apomorphous character a″ in A and B, or the apomorphous character b″ in A and C, arose by convergence or parallel evolution. In many cases we can decide with certainty which of these two possibilities corresponds with the factual course of phylogeny only through fossil finds.

The systematics of the Echinoidea furnishes a good example of this. The recent Echinoidea belong to three monophyletic groups, the Cicaridae (A), the Echinothuridae (B), and a third group (C) that includes all remaining families. Group C possesses several undoubtedly apomorphous characters. To determine whether any two of these groups (A, B, C) are closely related to each other we must investigate whether there are apomorphous characters in which A and B, A and C, or B and C agree. There is no evidence for assuming close phylogenetic relationship between the Cidaridae (A) and Echinothuridae (B). The groups B and C, on the contrary, have several apomorphous characters in common and consequently a close relationship between them is possible. But the groups A and C also have an apomorphous character in common—the rigid corona. In this character the Echinothuridae (B) are more strongly plesiomorphous than A and C; they are the only recent group of the Echinoidea with a flexible armor (corona). For this reason several recent authors (e.g., Moore) have suggested a sister-group relationship with the Echinothuridae and all other recent Echinoidea (A + C). In this case the apomorphous agreements between groups B and C would have had to arise by convergence. On the other hand, if we assume a sister-group relationship between the Cidaridae (A) and all other modern Echinoidea (B + C), then the immovable corona must have arisen by convergence in A and C. The paleontological evidence makes a decision between these two alternatives possible by showing that the rigid corona of the modern Cidaridae arose independently from a flexible corona. Consequently the rigidity of the armor in groups A and C rests on convergence or parallel evolution, and therefore the only apomorphous character in which they agree cannot be interpreted as synapomorphy. Thus no difficulties oppose the assumption of close relationship between groups B and C, which is based on agreement in several apomorphous characters.

The example of the Bibionomorpha (Fig. 43) reveals yet another way in which paleontology is important to phylogenetic systematics: the availability of fossil species from the Upper Lias shows that the two sister groups Phryneiformia and Bibioniformia-Fungivoriformia were already separate at this time. This fact is shown not by the occurrence of *Protorhyphus* (Fig. 43C), but solely by *Eoplecia* (Fig. 43F) if this genus is properly interpreted. All the bibionomorphid characters of *Protorhyphus* are plesiomorphous, and consequently the modern Phryneiformia and the Bibionoformia-Fungivoriformia could both be derived from it. Thus *Protorhyphus* shows the geological age of the whole group Bibionomorpha, but not the age of the two subgroups. *Eoplecia* lacks a discoidal cell, and thus has at least one apomorphous character in common with the Bibioniformia-Fungivoriformia. Consequently all modern species of the latter (but not the Phryneiformia) could be derived from *Eoplecia*. The existence of the group Bibioniformia-Fungivoriformia (as represented by *Eoplecia*) in the Upper Lias forces us to conclude that its sister group (the Phryneiformia) must have already existed at that time.

Contrary to a widely held opinion, it is not the plesiomorphous forms—the so-called collective types, from which derivation of many derived groups is formally possible—that are important in determining the age of monophyletic groups. Rather it is the forms with derived characters. This fact is of decisive importance in determining the absolute ranking of systematic groups. It will be discussed in detail in the next chapter.

Summary. The species is the given unit in the taxonomy of the higher group categories. The task of taxonomists working on higher taxa is to gather species into groups of higher rank according to the degree of their phylogenetic kinship. These "groups of higher rank" must be monophyletic. A monophyletic group is a group of species that arose by species cleavage, ultimately from a common stem species that is the stem species only of those species included in the group in question.

In contrast to the taxonomy of the lower taxa, the taxonomy of higher taxa is in the difficult position that the genetic relations between species cannot be determined directly. Consequently the determination is restricted to indirect methods. These were distinguished as paleontological, holomorphological, and chorological.

Superficial consideration might suggest that the paleontological method permits direct determination of phylogenetic relationships, but this is not true. It can be employed only in cooperation with the comparative holomorphological method.

The latter occupies a central position among the aids in the taxonomy of the higher taxa.

Taxonomy can begin its grouping task with the assumption that the degree of similarity between species corresponds with the degree of their phylogenetic kinship. This assumption can be tested, among other ways, with the aid of the phenomena of metamorphism and cyclomorphism—by treating the different stages of metamorphosis and cyclomorphosis as independent organisms, and then determining whether the systems derived independently for the different stages are congruent. Such tests show that the degree of morphological similarity—insofar as this can be measured at all—does not always agree with the degree of phylogenetic kinship. Nor is it possible to infer degree of phylogenetic kinship from the nature of the differential characters or corresponding characters. The true method of phylogenetic systematics is not to determine the degree of morphological correspondence, or distinguish between "essential" and "nonessential" characters, but to seek out synapomorphous correspondences.

The first step in finding synapomorphous correspondences is to identify, among different species (or species groups), those corresponding or divergent characters that belong to one and the same phylogenetic transformation series. This is the problem of homology of characters in a broadened sense. Then it must be determined which characters in the individual transformation series are to be considered plesiomorphous and which apomorphous. Only the mutual possession of apomorphous characters (synapomorphy) justifies the assumption that their bearers belong to the same monophyletic group. There are criteria and rules for determining whether the characters of different species belong to a transformation series, and for deciding whether they are to be evaluated as plesiomorphous or apomorphous, but these do not have absolute validity. Their applicability is restricted by reversibility, convergence, and parallelism in the phylogenetic development of characters.

The concept of the (holo)morphological similarity must thus be broken up, for the tasks of phylogenetic systematics, into the concepts symplesiomorphy, synapomorphy, and convergence (Fig. 44). In systems where, without this distinction, groupings are based on the simple principle of morphological "similarity" there arise group formations (corresponding to the complex nature of this concept) that may be distinguished as monophyletic (if the similarity is synapomorphy), paraphyletic (if the similarity is based on symplesiomorphy), and polyphyletic groups (if the similarity is due to convergence) (Fig. 45). While polyphyletic groups, where recognized as such, are today probably no longer permissable in any system, the distinction between paraphyletic and monophyletic groups is still often overlooked.

The paraphyletic groups (much as the polyphyletic ones) are distinguished from the monophyletic ones essentially by the fact that they have no independent history and thus possess neither reality nor individuality. They have no ancestor in common only to them, and thus also no point of origin in time in common only

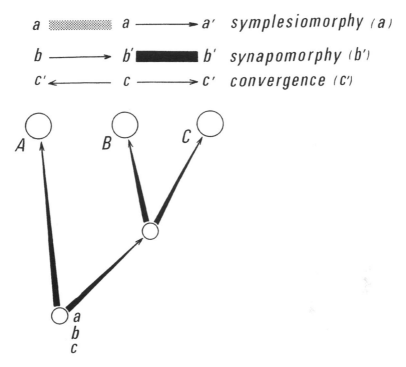

Figure 44. Diagram explaining the composite nature of the concept "similarity." Similarity can be based on symplesiomorphy, synapomorphy, or convergence.

to them in the true historical course of phylogeny. The concept of "extinction" has thus a different meaning for them than it does for monophyletic groups. If one uses a system that includes both paraphyletic and monophyletic groups as the basis for investigations concerning the gross course of phylogenesis and its conformities to law, faulty conclusions can hardly be avoided (cf. Hennig, 1962, on the chronological and epochal presentation of history in phylogenetic research).

The correctness of the results of taxonomic work is checked according to the principle that "true" is equivalent to "true to relationship" and "fruitful in the determination of new relationships." The criteria for the truth of the assertions of phylogenetic systematics are basically the same as those used in other sciences. Systematics investigates whether phylogenetic relationships are correctly inferred by means of a particular system into which the species and species groups are ordered on the basis of certain relationships existing between them. It investigates further whether relationships between the species other than those on which the system was based also have an orderly structure. Naturally such relationships must also have been brought about by phylogenesis if the testing is to make any sense.

The geographic distribution of organisms provides another way of checking the reliability of systematic results. Following the principle of "reciprocal clarification" it is possible in reverse order to use geographic distribution for determining the phylogenetic relationships themselves. In general, two taxonomic groups that stand in a spatial vicariance relationship to each other are more closely related than either is to any other taxonomic group. It has long been recognized that morphological similarity relationships have a reticular structure,

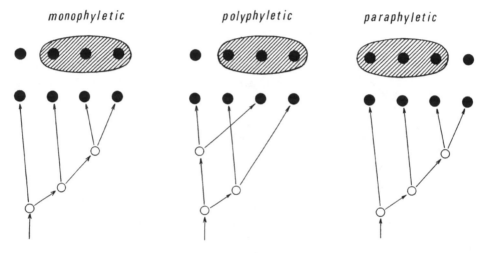

Figure 45. Group formations in typological systems. If species are ordered according to "similarity" without regard to the composite nature of this idea, then there are monophyletic groups (if the similarity is based on synapomorphy), polyphyletic groups (if similarity is based on convergence), and paraphyletic groups (if similarity is based on symplesiomorphy).

whereas the structure of phylogenetic relationships is hierarchic. The apparent contradiction that phylogenetic systematics tries to order species into a hierarchic system by analyzing relations that have a reticular structure is resolved when it is recognized that phylogenetic systematics does not use degree of morphological similarity in determining phylogenetic kinship relations.

Helmcke's treatment of the Brachiopoda (Fig. 46) shows clearly the different roles played by morphological correspondences and differences in a system that tries to express the morphological (typological) "relations" of species (expressed in the degree of their similarity), and in phylogenetic systematics (which is based on synapomorphy of the correspondences). Helmcke presents the relationships ("kinship" according to his own designation) of five families of brachiopods in

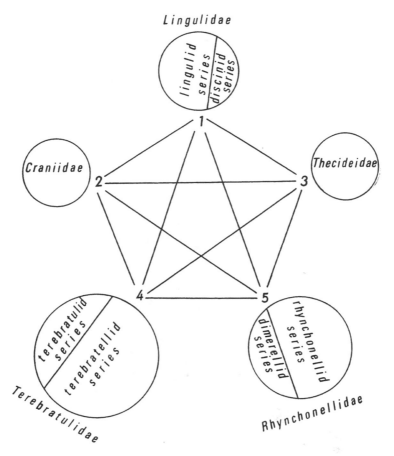

Figure 46. Schematic representation of the "kinship relationships" of five families of the Brachiopoda. (After Helmcke.)

the form of a reticular diagram. "Expressed in words, the relationships of the families are:

(A) 1-2   Central portion of shell cavity filled with organs. Intestine long, divided into a narrow forward loop and a posterior rectum, opening to the outside through an anus. Muscle bundles penetrated medially by transverse plate of connective tissue and also innervated there. Body wall with cutaneous muscle layer. Dorsal and ventral mantle lobes separated. Valves not articulated. Lophophores without supporting ridges on dorsal valve.

(B) 2-3   No pedicle.

(C) 4-5   Gonads present only in mantle lobes. Pedicle without cavity. Middle portion of muscles tendonized.

(D) 1-2-3   Muscle bundles fleshy, without tendinization.

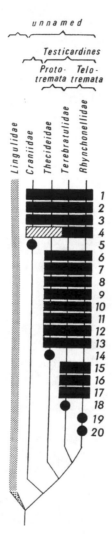

Figure 47. Phylogenetic kinship relationships of the five families of the Brachiopoda, for comparison with Fig. 46. The characters 1-20, which served as the basis of monophyletic groups, are the same as those on which the diagram in Fig. 46 was based. In each case only the apomorphous expression stages (black bars and circles) are registered. The diagram corresponds to the "scheme of argumentation of phylogenetic systematics" (Fig. 22). Only the "family Lingulidae" cannot be established as a monophyletic group on the basis of the characters given by Helmcke.

(E) 3-4-5   Central portion of shell cavity only partly filled with organs. Intestines ending in a blind sac. Dorsal and ventral mantle lobes united posteriorly. Valves articulated by a tooth and socket arrangement. Dorsal valve with ridges that support lophophores.

(F) 1-2-3-4 One pair of metanephridia. Mantle papillae present.
(G) 1-2-3-5 Right and left arms of lophophores separated.
(H) 1-2-4-5 Musculature barely reduced.
(I) 1-2-4-5 Dorsal mantle lobe completely fused with visceral sac.
(K) 2-3-4-5 Shell consisting of calcium carbonate. Anterior part of body parenchyma-
tous. Marginal lacuna not developed.

The reticular structure of the kinship relations here results from the fact that plesiomorphous and apomorphous stages of character expression are combined at random. Thus the character under (H) is probably exclusively plesiomorphous, that under (B) exclusively apomorphous. In other cases both plesiomorphous and apomorphous stages are used. For example, almost all of the characters mentioned under (A) are the plesiomorphous stages of the characters mentioned under (E). The structure of the phylogenetic relationships shown in Fig. 47 is based on a distinction between plesiomorphous and apomorphous characters.

(A) a "Central portion of shell cavity filled with organs" is apparently the plesiomorphous stage of the character listed under (E): "central portion of shell
cavity only partly filled with organs." This apomorphous condition is entered in the kinship diagram (6).

(A) b "Intestine long, divided into a narrow forward loop and a posterior rectum,
opening to the outside through an anus." This is apparently the plesiomorphous stage of the character listed under (E): "intestine ending in a blind
sac." This apomorphous stage of expression is entered in the kinship diagram (7).

(A) c "Muscle bundles penetrated medially by a transverse plate of connective
tissue and also innervated there." Helmcke does not mention the alternative stage of expression of this character; apparently it is absence of a
transverse plate of connective tissue. It is not clear which is the plesiomorphous and which the apomorphous stage. In view of the distribution
of the other characters, we would like to consider the presence of a transverse plate as part of the ground plan of the Brachiopoda and regard it
as a plesiomorphous character. Absence of the plate would then be entered as an apomorphous character in the kinship diagram (8). If it should
turn out that absence of the plate belongs to the brachiopod ground plan
and its presence is an apomorphous character, we would have to conclude
that the pedicle of the Terebratulidae and Rhychonellidae is not homologous to that of the Lingulidae (as Shrock and Twenhofel assume it to be).

Thus the presence of a pedicle in the lingulids, terebratulids, and rhyn-
chonellids could not be regarded as symplesiomorphy. The pedicle of the
Terebratulidae and Rhynchonellidae differs in various respects from that
of the Lingulidae: it is massive, contains no coelom invagination, is not
contractile, and is said to arise ontogenetically from the posterior part of
the larval body instead of from the ventral mantle lobes as in the Lingulidae. Consequently it is conceivable that the original "ground-plan pedicle"
of the Brachiopoda has been retained only in the lingulids, that it was

reduced in the further evolution of the brachiopods (this stage would be retained in the Craniidae and Thecideidae), and that still later a new pedicle differentiated on a different morphological basis (Terebratulidae and Rhynchonellidae). If this hypothesis is confirmed, the absence of the "ground-plan pedicle" would have to be entered in the relationship diagram as an apomorphous character of the Craniidae and Testicardinae, and the absence of a pedicle in the Craniidae and Thecideidae as a plesiomorphous character related to the presence of a new pedicle in the Terebratulidae and Rhynchonellidae. In any case this is one of the points that would have to be cleared up by future study.

(C) a "Gonads present only in mantle lobes." The alternative expression of this character is not given by Helmcke. Evidently it would be with the gonads lying partly in the metasome, which, in view of the position of the gonads in the Bryozoa, is to be regarded as the plesiomorphous stage. The fact that the gonads are restricted to the mantle lobes only in the Terebratulidae and Rhynchonellidae, which in many other respects are derived families, supports the view that this is a derived character (15 in the relationship diagram).

(C) b "Pedicle without a cavity." This is evidently the apomorphous stage of a character whose plesiomorphous stage has an evagination of the coelom within the pedicle. It is not clear whether the "pedicle cavity" disappeared in the Terebratulidae and Rhynchonellidae (entered under 16 in the relationship diagram) or whether they developed an entirely new pedicle that never had a cavity. The interpretation of the Thecideidae depends on the answer to this question. Since in this group the pedicle disappears during ontogeny, detailed study on the ontogeny might show whether it is a "pedicle with cavity" or a "pedicle without cavity" that is eliminated. This would help answer the question whether the pedicle of the Terebratulidae and Rhynchonellidae is or is not homologous with the pedicle of the Lingulidae.

(C) c "Middle portion of muscles tendinized." This is evidently the apomorphous stage of the character named under (D): "muscle bundles fleshly, without tendinization."

(D)   See under (C) c.

(E)   All characters listed here are apomorphous stages of the characters listed under (A) a-b and (A) e-g.

(F) a "One pair of metanephridia." This evidently represents the ground plan of the Brachiopoda, and is the plesiomorphous stage of an apomorphous character present only in the Rhynchonellidae: two pairs of metanephridia (entered under 19 in the relationship diagram).

(F) b "Mantle papillae," according to Helmcke, are "cells that fill the pores in the valves. Depending on the family, these mantle papillae are branched or tube-shaped." According to Helmcke, mantle papillae are absent only in the Rhynchonellidae. Consequently this could be considered an apomorphous character of the Rhynchonellidae.

Presence or absence of mantle papillae evidently corresponds to shell structures that, particularly in paleontology, are designated "punctate" and "nonpunctate." According to paleontologists (e.g., Moore) the punctate condition has repeatedly arisen independently among brachiopods. This is supported particularly by the data of Elliott (1953), according to which the Paleozoic Thecideidae have exclusively "pseudopunctate" shells, whereas in younger (Mesozoic and recent) forms this pseudopunctate structure is overlain by a punctate structure. Therefore we must assume that nonpunctate shell structure—and thus absence of mantle papillae— is a plesiomorphous character. In the relationship diagram the absence of mantle papillae in the Rhynchonellidae is entered with great reservation as an apomorphous character (20).

(G)   "Right and left arms of lophophores separated." This is evidently the plesiomorphous stage of a character that occurs in apomorphous condition (18 in the relationship diagram) only in the Terebratulidae: "each arm consisting of a semicircular plate in the most primitive genera. In the more complicated genera each arm is at first drawn out into a radius, and only then does it become wound into a spiral, which however is connected throughout its length with the spiral part of the other arm by a thin membrane" (Helmcke). From this we could infer that the apomorphous stage (the two arms connected) occurs only in the "complicated" genera of the Terebratulidae and is not part of the basic plan of this group. This question would have to be clarified.

(H)   "Musculature barely reduced." This is evidently the plesiomorphous stage of a character whose apomorphous stage occurs only in the Thecideidae: "musculature consisting of only three pairs of muscle bundles." These statements are somewhat incomplete, since the musculature is somewhat reduced in all Testicardinae—therefore in the Terebratulidae and Rhynchonellidae as well as in the Thecideidae. This is related to the presence of a shell lock in the Testicardinae that presents lateral movement of the valves. Reduction of the musculature is merely especially advanced in the Thecideidae (14 in the relationship diagram).

(I)   "Dorsal mantle lobes completely fused with visceral sac." This is evidently the plesiomorphous stage of a character whose apomorphous stage occurs only in the Craniidae: "the Craniidae differ from all other families in having the dorsal mantle lobes widely separated from the remaining visceral sac. There is a large cavity between the two organs into which seawater enters. The visceral sac and central cavity are connected only at the attachment sites of the muscles and the connection points of the mantle cavities" (5 in the relationship diagram).

(K) a   "Shell consisting of calcium carbonate." This is evidently the apomorphous stage of a character whose alternative (plesiomorphous) stage is found, according to Helmcke, only in the Lingulidae: shell composed of chitin with a minor amount of calcium phosphate. In the relationship diagram this apomorphous stage is listed under 1.

(K) b "Anterior part of body parenchymatous." This is evidently the apomorphous stage (listed under 2 in the relationship diagram) of a character whose plesiomorphous stage occurs in the Lingulidae: anterior part of body with cavities lined with coelomic epithelium.

(K) c "Marginal lacuna not developed." The alternative (plesiomorphous) stage of this character, which must occur only in the Lingulidae, is not described by Helmcke. It is not clear from his description what the marginal lacuna is (3 in the relationship diagram).

To these characters would have to be added that, according to Helmcke's data, a peculiar form of brood case occurs in the Thecideidae: "in the Thecideidae, the base of the arm of the oldest tentacle is transformed into organs of brood care." This character can also be entered in the relationship diagram (14).

From the above review it appears that only the Lingulidae are not yet confirmed as a monophyletic group; and that the Testicardinae, but not the Ecardinae, are monophyletic groups (Fig. 47). The Craniidae appear to be more closely related to the Testicardinae than are the Lingulidae. This assumption could scarcely be contradicted if it could be proved that the pedicle of the Terebratulidae and Rhynchonellidae is a new formation.

ABSOLUTE RANKING OF HIGHER TAXA

General. The methods discussed in the preceding chapters lead to a hierarchic system. Species are gathered into groups according to the degree of their phylogenetic kinship, and these groups are coordinated or subordinated to one another. This system expresses only the relative rank order of the groups. Systematics also attributes absolute designations to these groups—it speaks of genera, families, orders, etc. Consequently the question arises as to the criteria by which the absolute rank of a particular group is determined. It is strange that many systematists do not recognize that determining the absolute rank of systematic categories is a very serious and important problem. This is true even of systematists who accept phylogenetic systematics to the extent of demanding that the system must express phylogenetic relationships. They regard the question of whether a particular group is monophyletic or not as important, but consider it unimportant whether such a monophyletic group is called a genus or a family. Consequently they find that their genera, families, etc. are not comparable among the various phyla.

Naturally this is not correct. If systematics is to be a science it must bow to the self-evident requirement that objects to which the same label is given must be comparable in some way. Monophyletic groups belonging to different phyla actually are comparable in many respects: number of species, morphological differentiation, geological age, geographic distribution, position in the environment, etc. The question is, which of these criteria should be the basis for absolute rank order?

It has been proposed in all seriousness that number of species be used as a yardstick for determining generic rank: a genus should not contain more than about a dozen species. In present-day systematics the absolute rank of higher taxa is based almost entirely on analysis of morphological similarities: whether a species group is called a subgenus, genus, or family depends on the degree of morphological difference that separates this group from others. "Genera should include primarily all complexes of species that differ constantly from neighboring complexes in one or more fundamental characters" (Diels 1921). Mayr, Lindsay, and Usinger (1953) and Thompson (1956) are in essential agreement with this. The recommendations of Brauer (1885) have been particularly used for the higher taxa by theoreticians of systematic work. The works of Roux (1920) and Collier (1924), for example, are based on them: "If there are fundamental morphological, anatomical, etc. changes in homologous parts of the body in a series of families, then these may be elevated to orders." "If, in several orders, certain organs and organ systems agree in function, in essential morphology and anatomical characters, and in ontogenetic development, then these orders may be united into classes." Innumerable examples show that these principles are also followed in practical systematic work. The only objection made by Hendel (1928) to Handlirsch's proposal to unite the approximately fifty families of acalyptrates into a single family was that he was "unable to see in it the expression of progress in our phylogenetic insight, but only an obscuring of truly great differentiation of forms." Similarly, Schmitz (1939) says in the introduction to his monograph of the Palearctic Phoridae: "The family Termitoxeniidae erected by Wasmann (1901) is no doubt closely related phylogenetically to the Phoridae. It is a very old small twig (about thirty species are presently known) of the same branch of the Diptera that produced the great sister branch, the Phoridae. The question of whether these two twigs have diverged sufficiently to be recognized as separate families is still unsettled."

In principle there is nothing against determining the absolute rank of taxa on the basis of degree of morphological divergence, provided this is done within the limits set by phylogenetic systematics—that sister groups be coordinate, and thus have the same absolute rank.

The Acrania and Craniota are certainly sister groups. Consequently they are given the same rank in the system, and are (usually) designated subphyla. The Acrania include only about fourteen species, the Craniota tens of thousands distributed over several classes, orders, families, etc. The question now arises whether all the species of the Acrania together correspond to one class of the Craniota, to several orders, to an order with several families, or even to only a family. The fourteen species of the Acrania are generally included in one family, probably because of their slight morphological divergence. Of the two rank levels—family and subphylum—into which the monophyletic groups of these four-teen species are classified, the former is based on morphological divergences among the species themselves, the latter on its relationship to its sister group, the Craniota.

This way of determining the absolute rank of systematic categories does not in itself violate the basic principles of phylogenetic systematics so far developed, but it has severe limitations—the lack of any methods of measuring morphologi-cal differences. This has been discussed in great detail above. To see the point it is only necessary to face the problem of determining the degree of morpho-logical divergence between the cyclostomes and birds, or whether the morpho-logical divergence between an earthworm and a lion is more or less than that between a snail and a chimpanzee. It is often asserted that the Pentatomidae and Reduviidae, for example, are regarded as only families of insects, whereas the Rodentia and Lagomorpha (or even the Rodentia and Insectivora) are orders of mammals, because the differences between these mammalian "orders" are greater (have "ordinal" value) than those between the insect "families." I consider this sheer prejudice. In my opinion no one has ever presented even a shadow of proof that this gradation of differences actually exists. The difference in evaluation is based entirely on convention.

A further objection is that in groups in which there is a metamorphosis the divergence between species may be much greater or smaller in the larval than in the adult stage. There is no theoretically acceptable method of overcoming this difficulty, which has already been discussed in another context.

Finally, there is no reason why only morphological divergence should be con-sidered in the phylogenetic system. There may be very great differences between divergence in morphology and in other features such as behavior or psychic ca-pacities. It may be recalled that Huxley evaluates the psychic distance between man and other animals so highly that he suggests opposing man (as the repre-sentative of a special group, the "Psychozoa") to all other species of animals. But where is the yardstick by which psychic capacities can be measured, or by means of which they could be compared with morphological differences?

This example of the "Psychozoa" shows particularly clearly that determination of the absolute rank of systematic groups according to the degree of their diver-gence inevitably leads to a breakdown of the foundations of the phylogenetic system in all cases in which, usually as a result of accelerated character develop-ment, the divergence between a particular group and its sister group are dis-tinctly greater than the divergences ordinarily found in this particular super-ordinate group. Determination of absolute rank from (holo-)morphological divergence necessarily leads to ranking groups higher than their sister groups. This absolutely contradicts the fundamental construction principle of the phy-logenetic system, according to which sister groups must be coordinate. This was already clearly recognized by Naef: "It is an unspoken (unexpressed) convention of systematists to coordinate with the branches that arose from them those stem groups if they show a certain inner inclusiveness. But this is an unmistakable infringement on the principle on which the system is otherwise based." Naef agrees to this infringement and formulates the ground rule: "as systematic groups of equal rank may be coordinated (a) subgroups that arose by divergent cleavage of a type, or (b) a stem group and a group whose particular stem form or particular type differs in basic ways from the general type." Even today most systematists undoubtedly proceed according to these principles (see, for example, Mansfeld 1952). The naive way in which this is often justified is shown by an observation of Handlirsch (in Schröder, III) on the insect order Strepsiptera, whose sister group he seeks (no doubt erroneously) among the families of the malacodermous Coleoptera. "It would certainly be justifiable," he says, "to deal with the whole Strepsiptera, with all the countless suborders, superfamilies, etc. into which they have been split, simply as a family of the Malacoderma. I am not in favor of drawing such far-reaching conclusions from phylogeny, and simply ignoring the profound morphological differentiation. Ultimately every group has to descend from something, and the mammals cannot be considered simply as a family of reptiles, or even of flagellate Infusoria."

Naef's requirements need not be discussed here. It is enough to point out that they lead to a negation of phylogenetic systematics. The questions of why we insist on a phylogenetic systematics with a sound theoretical base and consistently executed, the extent to which other differently constructed systems are necessary, why we must reject—at least by intention—syncretistic systems, have all been answered in detail elsewhere. The view, represented in Germany particularly by Stammer (1959), that the different velocities of phylogenetic development have to be taken into account in determining the absolute rank of phylogenetic categories, and this in the construction of the phylogenetic system in general, have been rejected in principle because—like the impossible views of Alberti (1955) —they end up in distinguishing between "coordinate" and "subordinate" monophyly on Naef's principle, and result in a syncretic rather than a phylogenetic system.

We must remember again that our task in conferring a particular absolute taxonomic rank on the various groups of the phylogenetic system aims at making these groups comparable in a particular sense. We have concluded that determination of absolute rank according to degree of morphological divergence does not appear impossible in principle, but that in many cases it is not practicable and in other cases it would necessarily have to cut through the structure of the phylogenetic system because it leads to ascribing different ranks to sister groups. Beyond this there is the question of whether comparison between systematic groups of equal rank is fruitful at all if equal rank is based only on correspondence in degree of morphological divergence.

In his study of "biological progress," Franz (1935) tried to present examples of "historical retrogression accompanied by specialization." "Among groups that are known only or almost only from living representatives, the chameleons may be mentioned as lizard-like but highly differentiated animals with a limited distribution compared with the 'true lizards.'" But the chameleons cannot be opposed in this sense to the "true lizards," by which Franz probably means the Squamata of the usual system. Phylogenetically the chameleons represent only a side branch of the family Agamidae characterized by notable separate phylogeny (thus apomorphous characters). In the same study Franz says that "the reptiles have not developed the same flexibility as the mammals; in many groups they have regressed while the mammals have progressed." In both cases Franz compares groups that seem comparable to him because in the present system they are assigned the same rank (orders or classes). But in both cases equal rank has been achieved at the expense of the structural principles of the phylogenetic system. The system in which the Squamata and Chamaeleontida are coordinated as orders, and the Reptilia and Mammalia as classes, is not a phylogenetic system; it is a typological system. Franz' conclusions would be permissible only from a comparison of monophyletic groups—i.e., groups of a consistent phylogenetic system—and Franz' examples of the Lacertilia and Reptilia are not such groups. The mistake arising from this is, in the terminology of logic, the *metabasis eis allo genos*.

This mistake is often particularly evident in zoogeographic investigations, or more accurately, investigations of the importance to phylogeny of historic changes in the surface of the earth. Michaelsen (1935) demanded a strictly phylogenetic system with reference to the geographic distribution of organisms. He correctly sees that because of the inadequacy of our research methods "the unnatural remaining groups cannot be used for geographic and geological discussions." But it is very characteristic of the present position of systematics and the theoretical elaboration of its foundations that even he finds no fault if small side branches that during phylogeny differentiated strongly in morphological features (strongly "apomorphous" groups) are given the same rank as the totality of all the other species in that particular group, even though they are more closely related to one part of the species of the group than to others.

"There can be no doubt that the groups of the ancient genus *Acanthodrilus* of New Zealand, Australia, South Africa, Chile, and Central America, which have long been spatially separated, are much farther apart in terms of blood relationship or number of species that can be traced back to a common root than are the younger New Zealand groups *Maoridrilus, Plagiochaeta*, and *Neodrilus* of the New Zealand *Acanthodrilus* group from which they originated. These younger New Zealand groups are properly given the rank of genera, whereas the phylogenetically equal if not even higher valued *Acanthodrilus* groups of Australia, South Africa, Chile, and Central America are not because no obvious characters have developed among them on the basis of which they could be described diagnostically."

Apparently it does not occur to Michaelsen, or to most other present-day systematists, that a system of oligochaetes in which genera are distinguished on the basis of morphological differentiation—or probably more correctly on the basis of convenience of diagnosis—can be used for phylogenetic-zoogeographic investigation only by a specialist who, like Michaelsen himself, knows exactly the extent to which the actual generic relations are broken up by the generic limits employed. In other words, it can be used only by a specialist who knows exactly the consistent phylogenetic system as well as the typological system of the Oligochaeta.

In a study of the faunal relations of New Zealand, Australia, and South America I recognized and showed clearly that many of the most interesting questions cannot be answered simply, because even in a single group (such as the Diptera) systems of many different families that have been developed by different specialists would have to be compared. Such comparison is largely prevented because it is simply impossible to determine the extent to which the phylogenetic relationships between sister groups have been broken up in determining the rankings of groups such as genera.

Questions relating to the absolute rank—and thus the comparability—of systematic categories are of broad significance and occasionally have assumed an almost philosophical character. Thus Cuénot (1940) believed that "evolution cladique"—the origin of new phyla—ended at least six hundred million years ago. The youth of the earth is said to have passed. "The axis is dead. Evolution continues, of course, but only within the clades, whose summit alone is living." Woltereck (1940) also finds it "altogether inconceivable that new kinds of structural plans could branch off from the presently dominant types such as the vascular plants, the mammals, or the insects." In his opinion "the production of new somatic types and structural plans in both the plant and animal kingdom has terminated on the earth."

"Individual phyla age and become extinct, but new ones always take their place (Schindewolf 1950); yet the world of organisms as the highest order of life form is probably likewise condemned, after passing its peak, to 'age' and become extinct at some remote time unless man is capable of holding back this process by appropriate measures. The vertebrates as a whole may already have reached their peak" (Müller 1955).

The conclusions of Cuénot, Woltereck, and Müller are depreciated, however, by the fact that the categories of the typological system (in which new "phyla" can theoretically arise at any time) are not carefully enough distinguished from those of the phylogenetic system. In the phylogenetic system the origin of new "phyla" is by definition impossible because sister groups must have the same rank; any newly arising species has another species as its "sister group." Under certain circumstances we could speak of a new "phylum" only if the stem species from which our new species split off was itself the sole representative of a phylum of its own. But since in the system there is not a single phylum that is represented only by a single species, nothing could arise by the splitting of any living species that could be called a "new phylum" according to the concepts of the phylogenetic system. In reality the situation is even more complicated.

The so-called phyla of the animal kingdom (at least of the Metazoa) are predominantly monophyletic groups and as such are constituents of the phylogenetic system. Consequently in their rank the fact is expressed that in the phylogenetic system (in contrast to the typological) the absolute rank order cannot be independent of the age of the group, since in the phylogenetic system the coordination and subordination of groups is by definition set by their relative age of origin. At least roughly—i.e., in cases where no uniform measure is available for the absolute rank order with all groups of the system—very old groups must always have a high absolute rank. Consequently in the conclusions of Cuénot and Woltereck the logical error of the metabasis is present. At any rate they show the importance of questions of absolute rank hierarchy of systematic groups, and the erroneous conclusions that may result from failure to pay attention to their foundations.

It seems evident from the above that morphological divergence is not a suitable way of determining absolute rank order if all groups of the same rank are to be comparable in a theoretically incontestable way. But this does not mean that there is no way of achieving this goal. In the phylogenetic system there are actually groups that are unquestionably comparable ("equivalent") in an exactly determined sense, and in every case these must have the same absolute rank. These are the groups that stand in a sister-group relation to one another. The two subphyla Acrania and Craniota are unquestionably comparable in this sense, and their comparison opens up a multitude of extremely interesting and important questions. One could say that the comparison of sister groups, their (holo-) morphological and ecological breadth of differentiation, their systematic fine structure, their distribution, etc. is equivalent to investigating the question of whether there is conformity to law in phylogenesis. The most important reason why sister groups are comparable to such an unsurpassed degree is their origin from the same "root," in other words the fact that they began their development from the same initial conditions and passed through it from the same prerequisites. A part of these prerequisites is the simultaneity of their origin. They share this character with other groups of the system, whereas the same initial condition —which is realized in the same stem species—is a specific peculiarity of sister groups. No doubt the character of simultaneous origin suffices to make different groups of the system comparable in a very definite sense—even if they are not sister groups—and to make the comparison seem fruitful. Expressed in another way, establishing the absolute rank of systematic categories according to age of origin would mean solving the problem of making directly recognizable—by their absolute rank—which groups are in a definite sense comparable. This measure is even a logical augmentation of the fundamental structural principle of the phylogenetic system. As was fully discussed in the first part, this structural principle is based on the assumption that, in ordering monophyletic groups in the system according to relative rank, they are arranged according to their relative age of origin. Consequently it is logical to augment relative age of origin by a determination of absolute rank according to absolute age of origin. Obviously this kind of determination of the absolute rank of systematic categories can never lead to a breakdown of the basic structural principles of the phylogenetic system.

Before going into the objections that may be raised against our conclusions, we will discuss briefly the available methods for determining the absolute age of taxonomic groups.

Measurement of Absolute Ages of Higher Taxa. The same methods used in determining relative age are available for determining the absolute age of higher taxa. These methods are the paleontological, the comparative morphological, and the chorological. The significance of each of these methods in determining abso-

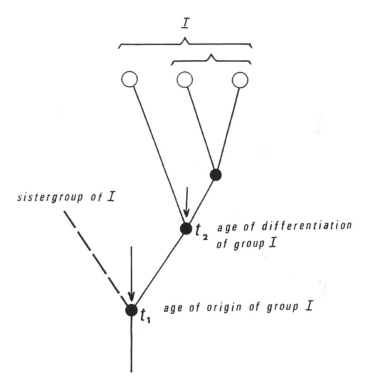

Figure 48. Diagram explaining the concept of "age of origin" and "age of differentiation."

lute age is very different from what it is in determining relative rank order. Of the chorological group of methods, practically only geographic distribution and the relations of broadly parasitic species groups to certain host groups are pertinent ("distribution in the environment"). In addition to the paleontological, the geographic and parasitological methods will be discussed.

A few remarks must first be made on the concept of the "absolute age" of an animal group. If we look at the phylogenetic tree of any group (Fig. 48), it is clear that two points in time are of decisive importance in the history of the group: the time when it separated from its sister group, and the time when the last common stem species of the recent groups ceased to exist as such. I have called these two points in time the age of origin (t_1) and the age of differentiation (t_2) (Hennig 1954).

From the point of view that naturally the age of differentiation of a group is identical with the age of origin of its oldest recent subgroups, the distinction between these two concepts may seem superfluous or even confusing at first glance. It is very important, however, especially in groups where the two points in time lie far apart. In such groups the temporal distance between t_1 and t_2 is often—though not always—expressed in greater morphological differences between the basic plan of all recent representatives of this group and its sister group.

If we study the literature on the age of animal groups we cannot avoid the impression that ignoring the difference between age of origin and age of differentiation has often led to confusing and sterile controversies. This is usually completely unnecessary because, in the phylogenetic system, groups whose age of origin and age of differentiation are very different have, so to speak, several different absolute ranks. Thus, to use an example employed above, the recent Acrania have at the same time the rank of a subphylum (corresponding to their age of origin) and a family (corresponding to their age of differentiation). At the same time they also represent a single class and a single order, to name only the so-called obligatory categories of the system. The importance of such "empty units" in the phylogenetic system has often been misunderstood. Naef, for example, contends that "if the number of 'nodes' of the phylogenetic tree is too small we will use 'empty' units—families with only one genus, genera with a single species, etc. Such units of course satisfy only the need for symmetry and surveyability of the system and for morphological juxtaposition, not the inner necessity of systematic summarizing."

These "empty units" are not meaningless in the phylogenetic system. The category of highest rank denotes the age of origin, that of lowest rank the age of differentiation of the group. Misunderstandings are impossible if there are different names for the different ranks of a group, with the aid of which the ranks can be unequivocally kept apart. Thus if we ask about the age of the Branchiostomidae it should be clear that we mean the age of differentiation of the Acrania, whereas a question about the age of the Acrania should refer only to the age of origin of this group.

In many cases there are no names that express at least roughly that the ages of origin and differentiation lie far apart. This is true for the fleas (Siphonaptera), for example. In these, in adaptation to a parasitic existence, particularly profound morphological changes took place between the time of cleavage from their sister group and the time of splitting up of the stem species from which all recent species arose. A question about the age of the Siphonaptera usually means: When did the stem species live that, as the first to live parasitically on birds or mammals, had the characteristic features—at least in their initial stages—of the basic plan of the Siphonaptera, and from which all recent fleas descended? Strictly speaking, even this species does not correspond exactly to the one we mean when we speak of the age of differentiation of the Siphonaptera, but it comes close to it in time. Paleontological data that would answer this question are lacking. With the aid of parasitological and geographic methods we would probably conclude that it was some time in the Mesozoic. But the fleas have characters that make it very improbable that they arose phylogenetically from one of the other "orders" of holometabolic insects. Therefore they can only be the sister group of one of these orders or a group of orders. Everything we know of their age indicates, however, that the Siphonaptera could scarcely have separated from their sister group later than the Permian. Probably the temporal distance between the age of origin and the age of differentiation of the Siphonaptera is even greater than was cautiously suggested above. Thus if we ask about the "age" of the Siphonaptera, we would have to ask whether the age of origin or the age of differentiation, or the first appearance of a certain combination of characters, is meant if misunderstandings and sterile controversies are to be avoided. This is likewise true in a great many other cases, and must always be kept in mind in the following discussion of the methods that serve for determining age.

The Paleontological Method. The paleontological method has the reputation of being the most direct and most reliable, if not the only method of determining the age of animal groups. But like all other methods it has its limitations. It fails in all groups that are incapable of fossilization, and to a considerable degree in those in which at best only superficial impressions of the body are preserved. It also fails for certain periods of the earth's history. Paleontology will probably never be able to give us the age of origin for animal groups that arose in the Archaicum of the earth's history. The record is very spotty even in strata that bear fossils. Zimmermann (1953) calculated that "on the average, of the several up to many billion ($= 10^{12}$) organisms of the past, in each case only a single one is available to science." For long periods of the earth's history modern animal groups are completely unrepresented by fossils, although we know from older accidental finds that they must have lived during those times. Consequently paleontology supplies at best only the minimal age, not the actual age, of an animal group. Only in very rare cases are fossil finds so numerous that separation of the stem lines of different groups can be adequately documented.

The significance of the paleontological method is further limited by the fact that fossils invariably present only portions of the "holomorphy" of an organism. Any systematist knows that the characters useful for determining phylogenetic relationships may lie in the most varied organs, sometimes in concealed organs, and sometimes only in certain ontogenetic stages (see also p. 34). The attempt to find them invariably in a particular organ is completely hopeless. This reduces

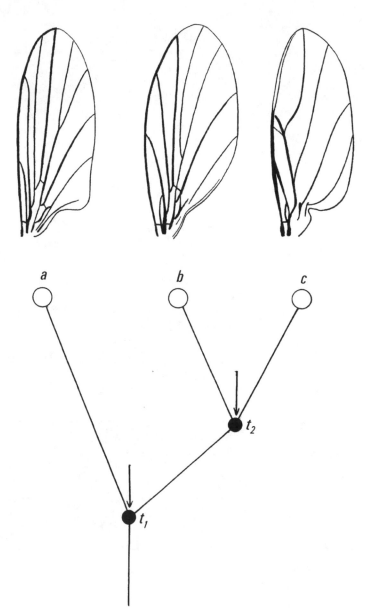

Figure 49. Kinship relations in three monophyletic dipteran families. a, Clythiidae, b, Sciadoceridae, and c, Phoridae. Compare with Figs. 50-52.

the value of many fossil finds. Naturally the expedient of creating a separate system for fossils "because the characters on which the neozoological system is based are not recognizable in fossils" is not useful for phylogenetic systematics, to which such independent paleontological systems are worthless. A much neglected problem, however, is to study those organs that are preserved in fossils in order to determine the extent to which they present characters from which conclusions about the relationships of the organisms could be drawn and that are as reliable as the characters that primarily led to the recognition of the organism. In many fossil insects, for example, the wings and their venation are the only characters that are recognizable with any exactness. But the wings do not always play the most important role in the phylogenetic system. A problem that so far has not been satisfactorily solved is the extent to which wing venation, in groups where it has been little studied, would permit recognizing kinship relations with certainty. Only after this has been done will it be possible to appraise properly the significance of many fossil finds.

All these, however, are so to speak only external circumstances that influence the value of the paleontological method for determining the age of animal groups. It is at least equally important that only fossils belonging to a particular group in the sense of phylogenetic systematics be used in determining the age of a group. This means that the fossil must be based on synapomorphous characters, that the species to which it belongs must truly be more closely related to all other species of the group in question than to any other known species.

A concrete example (Hennig 1954) will illustrate this. In Fig. 49 the relationships of three families of Diptera—the Clythiidae, Sciadoceridae, and Phoridae —are shown, together with the ground plan of their wing venation. It is evident that the three ground plans of wing venation are stages in a transformation series leading from the Clythiidae to the Phoridae. Assume that a wing known from the Upper Jurassic (time point t_x in Figs. 50, 51) corresponds with the wing of recent Clythiidae. The general practice is to include the fossil wing in the Clythiidae purely on the basis of such formal correspondence, and to assume that this proves the existence of the Clythiidae in the Late Jurassic. But wing venation in the Clythiidae is the most strongly plesiomorphous of the entire relationship group (Clythiidae + Sciadoceridae + Phoridae), so correspondence between the fossil wing and that of the recent Clythiidae is based on symplesiomorphy and does not prove any close relationship. Fig. 50 shows the possible interpretations of such a wing from time point t_x: it may really belong to the Clythiidae (A), it may belong to the common stem species of the entire group (B), to a pre-stage of this stem species (in which the characters of the wing venation have been reached, but not other peculiarities of the stem species) (C), or to the group Sciadoceridae + Phoridae, representing a stage at which those groups had already split off from the Clythiidae but had not yet changed the wing venation (D). Consequently such a find would tell us nothing at all concerning the temporal position

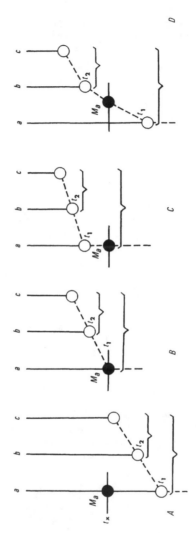

Figure 50. Significance of fossils for the age determination of animal groups. Assumed is a fossil (M_n) with the characters of the relatively plesiomorphous recent group a (compare Fig. 49) from the time t_x (e.g., Upper Jurassic). Such a fossil can make improbable the existence of any of the three groups a, b, and c at the time t_x.

of the points t_1 and t_2. Both points could lie before (A) or after (C) t_1 might coincide with t_x (B), t_1 might lie before t_x and t_2 after t_x (D).

The situation would be somewhat more favorable if a wing found at time point t_x corresponded with that of the recent Sciadoceridae, since venation is more

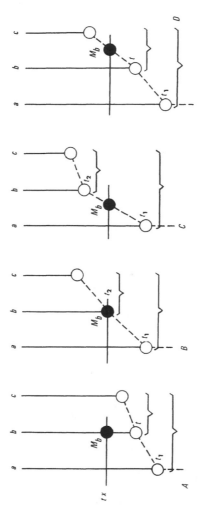

Figure 51. Significance of fossils for the age determination of animal groups. Assumed is a fossil (M_1) with the characters of the recent group b (compare Fig. 49) from the time t_x. Such a fossil determines the time point t_x as **terminus-post-quem-non** for the origin (t_1) of the sister-group relationships a— (b+c). It does not prove **for** the time point t_x the existence of group b (with which it formally agrees), but it does prove the existence of group c.

strongly apomorphous in the latter than in the Clythiidae. The resulting possibilities are shown in Fig. 51. Such a wing cannot be placed in the Sciadoceridae without further evidence, but it would permit us to determine the time point t_x (Upper Jurassic) as the *terminus-ante-quem* (not only as the *terminus-post-*

quem-non) for the time point t_1. The same would be true if a wing were discovered representing an intermediate stage between the Clythiidae and the Sciadoceridae. The most favorable find would be a wing corresponding with that of the Phoridae (Fig. 52). Except in case of convergence, such a wing could belong only to this group, since the Phoridae have the most strongly apomorphous venation of the whole group. It would permit us to establish the time point t_x as the *terminus-ante-quem* for t_1 and t_2.

These considerations show that fossils with only relatively plesiomorphous characters are of relatively little importance to systematics (Fig. 50), and that fossils with relatively apomorphous characters can be very important (Figs. 51, 52) because they not only prove the existence of the groups to which they belong but also may prove the simultaneous existence of other groups with strongly plesiomorphous characters. In the above example we would have to assume that if a wing of the Phoridae is discovered at the time t_x, then representatives of the more strongly plesiomorphous Sciadoceridae and Clythiidae must also have lived, not to mention representatives of even more strongly plesiomorphous groups. The conclusion—which seems paradoxical at first—is that favorable fossil finds not only say something about the existence of the group to which they belong, but just as much about the simultaneous existence of groups to which they definitely cannot belong. This is certainly not a new insight, but the possibilities that it offers have scarcely been exhausted. All this is true, of course, only if convergence or retrogressive development can be excluded. The procedure for doing this was discussed above (p. 116).

The example of the Clythiidae, Sciadoceridae, and Phoridae is purely hypo-

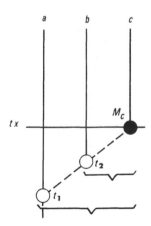

Figure 52. Significance of fossils for the age determination of animal groups. Assumed is a fossil (M_c) with the characters of the recent group c for the time t_x (compare Fig. 49). It determines the time point t_x as **terminus-post-quem-non** for the origin (t_1) of the sister-group relationships a—(b+c) and also (t_2) of the sister-group relationships b—c. It proves for the time point t_x not only the existence of the group c, but also the existence of the groups b and a.

thetical, since no fossils of these groups are known. However, the statements I have made (Hennig 1954) regarding the age of the most important groups of Diptera were based on the considerations demonstrated by this example. It might also be applied to conclusions regarding the age of the Mammalia. Many fossils from the Mesozoic are customarily placed among the reptiles because they do not have certain "essential" mammalian characters. Synapomorphous correspondences with the mammals show, however, that these particular fossils are more closely related to the mammals than to any group of the so-called reptiles. Consequently von Huene logically unites them with the mammals in a group Theromorpha. This corresponds with the basic principles of phylogenetic systematics. According to the view of most vertebrate systematists and paleontologists the fact that the older Theromorpha—contrary to basic principles of phylogenetic systematics—are almost universally placed among the "reptiles" is unimportant, probably because they believe the actual phylogenetic relations are so clear and so generally known that no misunderstandings can arise. It is quite otherwise in any case where the actual relationships are clear only to a narrow group of specialists. In Figs. 53 and 54 are two presentations on the development of the cephalopods taken from the same book. According to one (Fig. 53) we would have to assume that the Dibranchiata arose in the Devonian, and in the text it is stated that they are known as fossils since the Mississippian. From the other presentation it would seem that the two sister groups, the Tetrabranchiata to which among recent species only *Nautilus* belongs, and the Dibranchiata which includes all other recent cephalopods, have been separate since the Lower Ordovician.

The Biogeographic Method (Vicariance Type Doctrine). The biogeographic method starts from the observation that sister groups often differ in their geographic distribution—they often replace each other in different geographic areas. If the phenomenon of sister groups replacing each other in space is called vicariance, then we can speak of "vicariance relations of higher order" when dealing with sister groups of higher systematic rank. This distinction is necessary because the concept of vicariance was originally used in systematics only in connection with the discrimination of subspecies.

Vicariance relations of higher order need not necessarily be so pronounced that the ranges of sister groups are completely separated. They may partly or wholly overlap, resulting in a completely closed area for the group as a whole. It is decisive that we speak of vicariance relations only if a few low-ranking taxa of one or both sister groups occur in the area of overlap. We may assume that such cases represent secondary overlapping of ranges. The best known and most impressive vicariance types are those in which the ranges of sister groups are relatively widely separated on different continents. In zoogeography these are often called examples of disjunctive distribution or "disjunctions." Darlington (1957) showed thirteen such disjunctions on a map of the world. It is important that similar disjunctions may be found among very different animal groups, and this raises

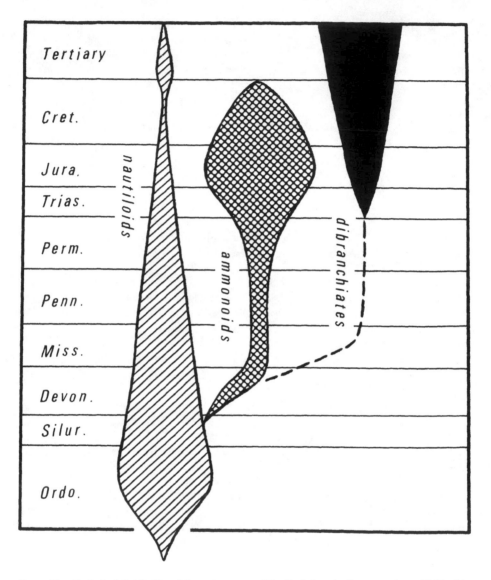

Figure 53. Geological distribution of the main groups of the Cephalopoda. For comparison with Fig. 54.

the question of whether they arose in the same or a similar way in all animal groups. If so, this might mean that the same age (more accurately, the same age of differentiation) would have to be attributed to all groups showing the same or a very similar picture of disjunctive distribution.

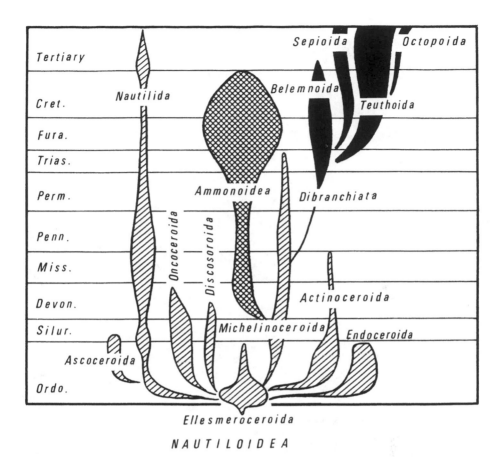

Figure 54. Geological distribution of the main groups of the Cephalopoda. For comparison with Fig. 53. Both diagrams from Moore, Lalicker and Fischer, Invertebrate Fossils. Copyright 1951 by McGraw-Hill Book Company.

This question is best investigated in groups in which one of the partial areas lies on one of the peripheral continents such as South America or Australia. For these peripheral or island continents, but also for other relatively closed geographic areas, faunal layering and the ages of the various faunal layers have been frequently investigated. Such investigations have sometimes been discredited by uncritical attempts to distinguish a great number of faunal levels of varying origins and ages, and to reconnect these by hypothetical land bridges postulated on very inadequate grounds. This does not mean that such studies should simply be abandoned. Especially for South America and Australia it can be shown that results that at first sight seem to show little agreement are not so different that broad agreement in a few essential points cannot be demonstrated (Hennig 1960).

It is now rather generally assumed that South America was connected only with North America during the Cenozoic, and that this connection was interrupted for a long period in the early Tertiary. "After the Paleocene, at least, there was no real interchange until well along in the Miocene" (Simpson 1953). Corresponding to these two land connections prior to and after the interruption, two faunal layers are distinguished: "before or at the beginning of the Tertiary, immigrants representing some but not all contemporary Old World stocks of cold-blooded vertebrates somehow reached South America and persisted and radiated there. And toward the end of the Tertiary additional immigrants . . . reached South America across the existing land bridge" (Darlington 1948). Arldt (1938) called the two faunal layers the "Edentate Layer" (uppermost Cretaceous and possibly Paleocene) and the "Felid Layer." Unfortunately his attempt to assign numerous animal groups to one or the other of these two layers was so uncritical that it is practically worthless. Dunn (1931), Darlington (1948), and Simpson (1943) have done this far more critically for the amphibians, reptiles, and mammals. The basic importance of such work to determination of the age of animal groups is obvious. In many cases referral to one of the two faunal layers is assured by the occurrence of the first fossils in South America. The minimal age of such groups would then be determined solely by the fossils. In other cases no fossils are known and allocation to one of the two layers is based on other considerations.

Allocation of a group to the older faunal layer (Arldt's Edentate Layer) carries with it an assumption as to the minimal age of the group. For example, Darlington (1948, 1957) says that the Pipidae "arrived early" in South America, which puts them in the Edentate Layer. Since arrival in South America no later than the Paleocene is part of the definition of the Edentate Layer, we are also making the assumption that the sister-group relationship between the South American Pipidae (*Pipa*) and the African Pipidae (*Xenopus* and two other genera) dates from no later than the upper boundary of the Paleocene. Thus if there were safe criteria for assigning animal groups to the various faunal layers of South America and other continents, we could compensate to some extent for the complete absence of fossils of so many groups. Makerras presented such criteria for the Diptera of Australia (see also Hennig 1960). Naturally, there are no criteria that apply in the same way to all animal groups. For very vagile groups it would scarcely be possible to draw safe boundaries between faunal layers of different age. Like all other methods, the biogeographic method has limited possibilities of use. It seems, however, that with very cautious evaluation of criteria it should be possible in most cases to decide whether a particular group arrived in South America or Australia before or after the end of the Upper Oligocene. This point in time would then be the *terminus-post-quem-non* for the origin of sister-group relationships among animals that immigrated in the early Tertiary and their living sister groups living outside one of these two continents.

It is questionable whether there are also points in time in the Mesozoic that could be used with some certainty for determining minimal ages of animal groups or sister-group relationships. For the mammals of Australia, Simpson (1953) distinguishes "archaic immigrants"—which he regards as Late Triassic or Jurassic—in addition to the more recent faunal layers of the Late Cretaceous or Tertiary. Simpson's archaic immigrants correspond to Arldt's "Monotreme Layer." In South America there were mammals—which left no living representatives—earlier than the Edentate Layer. Arldt refers them to his "Sparassodont Layer"; according to his interpretation they reached South America from Africa in the Mesozoic. Dating this Mesozoic faunal layer is tied up with questions regarding trans-Atlantic land bridges between Africa and South America. I believe it is impossible at present to say more about the oldest faunal layers of Australia and South America than that the boundary between the Upper and Lower Cretaceous is the *terminus-post-quem-non* for the origin of the oldest monophyletic groups on both continents. This agrees approximately with the view of Darlington (1957) that a safe correlation between geographic distribution and the history of recent animal groups cannot be established beyond the Cretaceous. The view of Jennel (1950) that (for insects) fossils permit reconstructing the history of major lines in the Paleozoic and Triassic, and that from the Jurassic on the same is true of biogeography, is probably too optimistic although not basically very different.

All this applies first of all only to animal groups of the continents and of fresh waters. Beyond question there are also vicariance relationships of higher order in marine groups. Good examples may be found among the corals. Thus according to Kükenthal (1919) the *Semperina* group of the Briareidae (Octocorallia-Gorgonariae) is restricted to the Indo-Pacific, the *Briareum* group to the Caribbean, with the North Atlantic genera *Anthotheria* and *Paragorgia* lying farther apart. The situation is similar for the Plexauridae and Muriceidae. Vicariance relationships are expressed most beautifully in the Echinoptilidae and Renillidae (Fig. 55). According to Kükenthal and Broch (1919) these represent a special side branch of the Pennatulaceae. The Echinoptilidae (3 species of the genera *Echinoptilum* and *Actinoptilum*) are distributed in the Indo–West Pacific, the Renillidae (one species of *Renilla*) in American waters. Parallelizing the vicariance types of the sea with those of the continents, once this has become possible, should be facilitated by the fact that the littoral faunas of the sea—to which the majority of the most important marine organisms belong—are closely connected to the continental blocks. This should make it possible to compare some of the major centers of distribution of marine groups with those of the terrestrial organisms. So far, however, we lack the foundation for determining the minimal age of vicariance, and thus of the sister-group relationships, of marine animal groups.

Even the sessile littoral forms have great powers of dispersal because of their pelagic larval stages. This, together with the continuity of all parts of the sea, give greater importance to ecological barriers to dispersal than to the mechanical-geographical barriers that are important on land. According to Ekman it is not so much the regionally bounded parts of the sea as the types of water that form

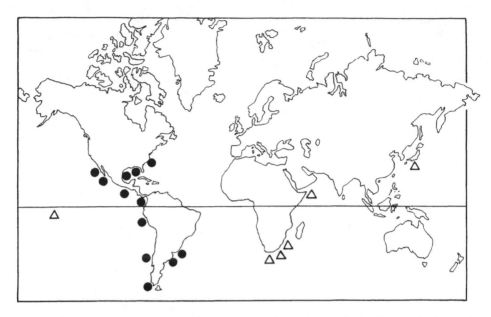

Figure 55. Geographic distribution of the coral (Pennatulacea) families Echinoptilidae (triangles) and Renillidae (circles). (Redrawn from Kükenthal and Broch.)

the elements of marine zoogeography. As a result, vicariance in the ecological realm may be expected to play a larger role in the taxonomy of marine organisms than it does among land forms.

The Parasitological Method. The paleontological and geographic methods can potentially be applied to all groups of organisms, and their applicability is limited only by the quantity of data and the possibilities of interpreting these data. The parasitological method, however, can be used only for certain groups of organisms that are connected with others by a host-parasite relationship.

For our purposes the concept of parasitism can be extremely broad, including forms not generally regarded as parasitic, such as the phytophages, the gall formers, and the mining insects. The relations between parasites in this broad sense and their hosts can be interpreted as a special form of chorological relationship, i.e., a relationship between the organism and its environment. Thus the parasitological method can be regarded as a component of the chorological group of methods and might then be compared with the geographic method. This view is no doubt correct, but it is particularly significant in the use of chorological methods for determining the relative rank order—i.e., relative degree of phylogenetic relationship—of taxonomic groups. As shown above, all other chorological relationships of organisms can actually be used for this too. It is particularly characteristic, however, that only a very special selection of chorological relations can be used, namely those between the organism and its environment as a food source. Not even all of the latter are usable, but only those where certain groups of higher or lower rank are restricted to other very definite groups of organisms as their source of food. Consequently it is probably justifiable to single out a particular parasitological method of taxonomy. There probably is no group of organisms that either does not have parasites or is not itself parasitic, so the applicability of the parasitological method is not as limited as might appear at first sight. It is always a prerequisite, however, that the absolute rank of one of the two members of a host-parasite relationship has already been determined by other means.

Evaluation of host-parasite relations for determining the absolute rank of one of the partners rests on the very general proviso that the two groups developed in parallel, or that one remained constant while the other split up and evolved further. The relations of the latter must not have been extended to organisms other than the successors of its original partner. Starting from these provisos, the occurrence of identical or at least very similar (and therefore very probably related) parasites on different hosts whose interrelationships are not so apparent has been considered proof of close relationship among the hosts.

All this was discussed in great detail above (p. 107), where we mentioned the difficulties and false inferences that may arise. There are additional possibilities for false inferences if use of the parasitological method is extended to determining the age of parasite groups. We cannot, for example, infer the age of parasites directly from the age of the host group. The bat flies (Nycteribiidae-Streblidae) are an extraordinarily peculiar group of Diptera that live exclusively as parasites on bats, but this does not mean that they have the same geological age as the bats. The stem species of the Nycteribiidae-Streblidae could very well have evolved anywhere in the world as an adaptation to parasitizing a particular species of bat, long after the bats themselves had split up into several narrow communities of descent. Once the adaptation was established, the descendants of the stem species could have invaded other species of bats but found suitable living conditions only on this one order of mammals. Consequently the occurrence of the Nycteribiidae-Streblidae on bats and their restriction to bats does not prove that the bats and bat flies arose together and therefore should have the same taxonomic rank.

Everything that was said above (p. 162) regarding the importance of the distinction between the age of differentiation and the age of origin of animal groups applies in a case of this kind. The restriction of the Nycteribiidae-Streblidae to bats provides a *terminus-ante-quem-non* for the age of differentiation of the bat flies. It is extremely probable that all recent species of Nycteribiidae-Streblidae are to be derived from a single stem species that already had all the adaptive characters belonging to the ground plan of this group. This stem species cannot have lived prior to the origin of the host group to which it was adapted, but it need not have arisen simultaneously with the host group. Thus the age of the bats (Chiroptera) tells us nothing about the age of origin of the bat flies, i.e., the point in time when the Nycteribiidae-Streblidae separated from their sister group (unfortunately the sister group is not known with certainty). This point in time may equally well have been before or after the origin of the Chiroptera.

Even if two host groups having a sister-group relationship are infested by two different groups of parasites that in turn are sister groups, it cannot be inferred that parallel evolution with simultaneous origin of the two sister-group relationships took place. According to Hering (1960), *Pegomyia hyoscyami* occurs on the Chenopodiaceae and Caryophyllaceae as two "biological races." Although the two groups of host plants are related (Centrospermae), and the differentiation of the two races of *Pegomyia hyoscyami* corresponds to the kinship relations of the host plants, we probably cannot assume that the races of *Pegomyia hyoscyami* arose at the time when the two plant families split from their common stem form of the Centrospermae. More likely the parasite first adapted itself to one of the families of host plants, and then succeeded in jumping the barrier that separated it from the other family, thereby evolving a new race. We can easily imagine that after a few million years each of the races of *Pegomyia hyoscami* could have evolved further into a complex of species groups, with one of the two sister groups still restricted to the Chenopodiaceae and the other to the Caryophyllaceae. Naturally, even then we could not infer—any more than we can today—that the two sister groups arose simultaneously and in parallel with the separation of the two groups of host plants.

Perhaps it is best to show by means of a concrete but clear example the considerations that are necessary in using the parasitological method if false inferences are to be avoided. Fig. 56 shows how Rubtzov (1939) interpreted the

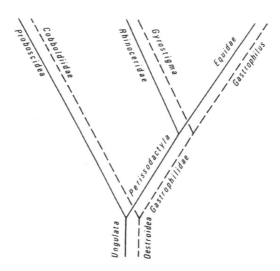

Figure 56. Phylogenetic parallel development of the Gastrophilidae (Diptera: parasites) and the Ungulata (hosts) following the interpretation of Rubtzov. Compare with Fig. 57. (Drawn after Rubtzov 1939.)

connection between the dipteran family Gasterophilidae and certain ungulate groups as a consequence of parallel evolution. The genus *Cobboldia* is restricted to the Proboscidea, *Gasterophilus* to the Equidae, and *Gyrostigma* to the Rhinocerotidae. At first sight the hypothesis of parallel evolution appears to explain this distribution best, but it has some blemishes. In the first place, it has not been shown beyond doubt that the Gasterophilidae are actually a monophyletic group; for example, many authors question a close relationship between *Cobboldia* and the other two genera. I consider it probable that the three genera do form a monophyletic group, but even so there is no unequivocal proof that each genus is itself a monophyletic group, and if this is true, that *Gyrostigma* and *Gasterophilus* are sister groups that, in turn, together form a sister group vis-à-vis *Cobboldia*. All this must first be established according to the "argument scheme of phylogenetic systematics" (p. 91). But even if this could be done—and there seems little doubt that basically it could—serious questions arise because all essential groups of the Gasterophilidae are entered in Rubtzov's diagram, but not all groups of the Ungulata. If Romer's family tree of the Mammalia—from

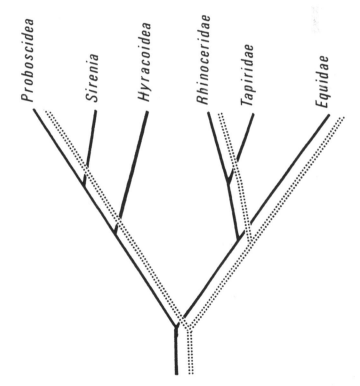

Figure 57. Phylogenetic tree of the Ungulata (solid lines, after Romer) and the Gastrophilidae (dotted lines) under the assumption of phylogenetic parallel development. Compare Fig. 56.

which Fig. 57 was taken—is correct, then there is no longer reason for assuming that the Proboscidea are more closely related to the Perissodactyla than are the Artiodactyla. The hypothesis of parallel evolution must reckon with the fact that it assumes a secondary disappearance of the Gasterophilidae from the Hyracoidea, Sirenia, Artiodactyla, and Tapiridae, since gastrophilids must have been present among the ancestors of all these groups (Fig. 57). To avoid this difficulty, what is to prevent us from assuming that the stem species of the recent Gasterophilidae became adapted to the Equidae after the latter were already a separate group, in other words at a time when a sister-group relationship already existed between the Hippomorpha (Equidae) and the Ceratomorpha (Tapiridae + Rhinocerotidae)? Descendants of this stem species could then have first passed over to the Proboscidea (ancestors of *Cobboldia*) and later to the Rhinocerotidae (ancestors of *Gyrostigma*), while some (immediate ancestors of *Gasterophilus*) remained on the Equidae. Such a hypothesis would be in extreme contrast to that of a parallel development since the time of origin of the ungulates, and perhaps is no more probable than the latter. There are other less extreme possibilities, but it is scarcely possible to choose among them at present. Consequently I do not think we can say more at present than that the time of origin of the ungulates (the uppermost Cretaceous, according to Romer) must be considered the *terminus-ante-quem-non* for the age of differentiation of the Gasterophilidae.

A more exact dating would be possible if all or almost all of the important ungulate groups were parasitized by Gasterophilidae, particularly by different subgroups of this family, and if furthermore the relative rank order of the sister-group relationships among the parasites corresponded exactly with those of the host groups. Then there would be no objections to a hypothesis of parallel development. The more richly differentiated a parasite group is, and the more accurately this differentiation corresponds to that of the host group, the more reliably the parasitological method will work.

Several cases will be cited here as examples. Szidat (1940) believes that among the trematodes he has succeeded in making "a classification of the families of trematodes into subfamilies and genera on the basis of their main hosts that directly parallels the classification of the trematodes on the basis of striking morphological and developmental characters." Thus according to Szidat the Fasciolopsidae, in which he was successful in such parallelizing, are to be derived from an echinostomous primary form that was already parasitic on creodonts. This primary form "had, like its hosts, split up into several developmental series that occurred as Campulidae in pinnipeds and whales or as Fasciolopsidae in the various groups of ungulates."

According to Fuhrmann (1908), the differentiation of the cestodes is also "closely tied to the appearance of the vertebrates." It is possible to make some estimate as to the time of origin and absolute rank of the opalinids (Protozoa) and their subgroups on the basis of their linkage to the Anura as hosts.

One cannot be too cautious in establishing that host and parasite groups developed in parallel. The fact that the "more primitive" or "older" subgroups of a parasite group occur in the "more primitive" or "older" host groups and vice versa is not in itself an adequate basis for such a hypothesis. There are cases where all this is true, although "parallel evolution" of parasites and hosts can be excluded with certainty. Among the Strepsiptera, for example, there appears to be a sister-group relationship between the Mengeidae and Stylopidae. The Mengeidae are characterized by a number of plesiomorphous characters and so far as known occur only in the Lepismatidae (Thysanura), whereas the Stylopidae have exclusively pterygotous insects as hosts. Some smaller subfamilies of the Stylopidae (regarded as families by many authors) occur in Orthoptera and Hemiptera, while the main mass is restricted to the Aculeata, a highly specialized and certainly relatively young subgroup of the Hymenoptera. Nevertheless we cannot speak of a parallel evolution between the Strepsiptera and their hosts, since the origin of the sister-group relationship between the primarily wingless "Thysanura" (Lepismatidae) and the Pterygota undoubtedly lies much farther back than the origin of the Strepsiptera, which themselves are only a subgroup of the Pterygota-Holometabola. Also, the sister-group relationships among the hosts of the Stylopidae must have arisen long before the Stylopidae themselves.

The situation is similar in the Tachinidae (Diptera). "The higher Tachinidae are typically parasites of Lepidoptera, whereas the most primitive of the lower Tachinidae . . . are parasites of beetles or forms with an originally unspecialized host selection which enabled closely related genera or groups often to become regular parasites of unrelated groups of hosts (Chilopoda, Dermaptera, Saltatoria, Embioptera, Coleoptera, Hymenoptera, Lepidoptera, Diptera, and the majority of the tribes in Phasiinae of Hemiptera Heteroptera)" (van Emden 1959). Here only the "most primitive" and "older" parasitic groups occur among the "more primitive" and "older" hosts, and the derived and younger parasites are characteristic of the relatively young and derived Lepidoptera. Nevertheless there can be no parallelism in the evolution of hosts and parasites because the Tachinidae, as a relatively subordinate subgroup of the Cyclorrhapha, are undobutedly much younger than the Lepidoptera.

Particularly favorable results can be expected if the parasitological method is supported by geographic vicariance relationships. The genus *Struthiolipeurus* (Phthiraptera) occurs only in ethiopian (*Struthio*) and neotropical (*Rhea, Pterochemis*) ratite birds. If *Struthiolipeurus* is really a monophyletic group, this would be a very strong argument for a close relationship between the host groups —a point often debated by ornithologists. This argument is supported by the choice of hosts of the tribe Deletrocephaleae of the subfamily Strongylinae (Nematoda). This tribe includes two genera: "*Codiostomum*, with a single species found in three species of ostriches in Africa, and *Deletrocephalus*, with a single species occurring in *Rhea americana* of South America" (Harrison 1928).

On the other hand we can infer the minimum age of the genus *Struthiolipeurus* and the tribe Deletrocephaleae from the vicariance type to which the

Struthionidae-Rheidae belong. The South American Rheidae can certainly be referred to one of the distinguishable faunal layers of that continent. Arldt (1938) assigns them to the Edentate Layer. This would mean that their ancestors reached South America no later than the Paleocene, and that there could have been no ecological contact since then between the sister groups, Struthionidae and Rheidae. But this also means that the separation between the neotropical and ethiopian species of *Struthiolipeurus* and between the two genera *Codiostomum* and *Deletrocephalus* must have existed at least since the end of the Paleocene. Consequently this point in time is undoubtedly the *terminus-post-quem-non* not only for the age of origin but also for the age of differentiation of *Struthiolipeurus*, and it has the same significance for the ages of origin and differentiation of the nematode tribe Deletrocephaleae (in which the "age of differentiation" coincides with the age of origin of the genera *Codiostomum* and *Deletrocephalus*).

Perhaps even better based is another example from the systematics of the Phthiraptera (Vanzolini and Guimaraes 1955).

South American marsupials are parasitized solely by lice of the genus *Cummingsia* (Mallophaga, Trimenoponidae.) Two of the three known species of *Cummingsia* parasitize didelphids. The third lives on a caenolestid. It is well established that didelphids and caenolestids were already well differentiated in the Paleocene, and even then their taxonomic relationships were not close. . . . Other trimenoponids occur on South American hystricomorphs, one species parasitizing secondarily one hare. No trimenoponids are known outside of South America. . . . The geographical distribution of the trimenoponids points to their origin on the marsupials and to subsequent contamination of the rodents. . . . On the family level, the closest relatives of the Trimenoponidae are the Boopidae, parasites of Australian marsupials.

If the Trimenoponidae and Boopidae are really sister groups, as they would appear to be, then in this case too the upper boundary of the Paleocene would have to be considered the *terminus-post-quem-non* for the origin of this sister-group relationship. Actually there has probably been no contact between the direct ancestors of the neotropical and Australian marsupials since the Upper Cretaceous.

The main difficulty in the way of a general use of the parasitological method is that only a minute amount of the possible knowledge of parasites and their hosts is available. Many fortunate circumstances must coincide and much difficult work be done to make known a single host-parasite relationship, and the discovery of even several new relationships scarcely promises anything very interesting. Only the comparative investigation as nearly as possible of all existing relationships will reveal the conformities to law. We cannot foresee the role that the parasitological method will play some day when relationships are more accurately known, particularly among groups living in non-European countries. We can be sure that it will not be small.

The Holomorphological Method. We have repeatedly emphasized that the methods discussed above for determining the absolute rank of higher taxa do not work in all cases. Each has a certain area of applicability where it works best. In some cases two or all three of the methods—paleontological, geographic, and parasitological—can be used and the results of each checked against the others. But there are cases in which none of these methods is applicable: the parasitological method is limited to parasitic and parasitized groups, the chorological (geographic) method fails with relict groups and groups whose origin goes back beyond a certain time, and the paleontological method is tied to the availability of fossils. So far the paleontological method has failed particularly—except in numerous individual cases—in groups whose origin lies earlier than the Paleozoic (Cambrian).

In all these cases the holomorphological method must be used in a way very similar to its use in determining the *relative* rank order of taxonomic groups. The problem is simplified by the fact that the other methods have already supplied a firm framework for an absolute rank order: in all groups of the most varied category status there are subgroups whose absolute rank can be determined beyond question with the aid of the paleontological, geographic, or parasitological methods. Consequently there remains for the holomorphological method merely to determine, so to speak by inter- and extrapolation, the absolute rank of those groups for which the other methods do not work.

The justification for this is derived theoretically from the observation that there seem to be evolutionary rates that are characteristic of entire animal groups ("horotely" of Simpson 1944). Strong deviations from this (bradytely and particularly tachytely) are the main reasons leading consciously or otherwise to a breakdown of the axiom that in phylogenetic systematics sister groups must have the same absolute rank. Naturally the evolutionary rate of a group is expressed in the morphological divergence among the species united in the group. Consequently it is morphological divergence that is used in employing the holomorphological method for estimating the age of a group. Thus we seem to have returned to a principle that was emphatically rejected above (p. 155), but this is only apparently true. If, for example, we put the fourteen species of Acrania in a single family on the basis of degree of morphological divergence, we could do so because the degree of morphological divergence serves us as a yardstick for determining the absolute rank of systematic categories. This is what we rejected. But we can also regard the particular degree of morphological divergence as expressing a certain evolutionary rate that led within a certain time to the observed degree of morphological divergence. We placed the fourteen species of the Acrania in one family because relatively little time seems to have passed (as may be inferred, in the absence of other indications, from the minor degree of their morphological divergence) since the existence of their last common stem form. This is the only case in which the holomorphological method of determining the absolute rank of a category appears to be theoretically comparable with the principle that absolute rank is based on age of origin. That this is not merely a theoretical subtlety is shown by the fact that if the evolutionary rate of a group seems to be much faster than the norm we take this into account and set the absolute rank correspondingly lower. In evaluating morphological divergence as such, however, groups with high divergence must be given a correspondingly high rank (see also von Wahlert 1957) regardless of how the great divergence came about: by a long period of evolution (i.e., great age of the group), or in a relatively short time by tachytely. As mentioned above, this leads to a breakdown of the principles of the phylogenetic system.

What we have said of morphological divergence also applies completely to numbers of species in taxonomic categories, which may also be regarded as the result of a certain rate of evolution. Number of species can rarely be used alone, and even then only as a very rough yardstick for measuring the age—and thus the absolute rank—of a systematic group. We might suspect merely from numbers of species—even if no other indications were available—that the dipteran "families" Limoniidae (over 5,000 species), Tabanidae (over 3,000 species), Asilidae (over 4,000 species), Empidae (about 3,000 species), Muscidae (over 4,000 species) are older and therefore to be given a higher rank than the so-called "families" of the Acalyptratae (about 150 species). The connection with other methods is also particularly promising here. For example, on the basis of number of species (about 200) alone we can probably rule out the possibility that the dipteran family Richardiidae belongs to the Felid Layer (see p. 172): monophyletic groups with large numbers of species cannot be very young. Naturally, the opposite conclusion, that groups with few species must be young, is not permissible. It is generally, and probably correctly, assumed that groups whose species are clearly separated morphologically are older than those in which the species can be distinguished only with great difficulty. Besides the progress of the true differentiation process, the extinction of intermediate forms—which becomes increasingly apparent after long periods of evolution—seems to be responsible for this. If this point of view is kept in mind we will often be saved from assigning too young an age to groups with relatively few species.

Summary. There is no single method with which the age of origin of systematic groups can be determined accurately and certainly. Even in the most favorable cases only minimal and maximal limits can be recognized. Consequently objections to determining the absolute rank of systematic groups on the basis of their age seem to be justified. The idea of determining the rank of a group by its age has hitherto been rarely expressed; it has always been rejected either by the author himself or by other critics.

Simpson (1937), for example, recognizes that some mammalian families (e.g., the Didelphidae) originated in the Cretaceous, others (e.g., the Ursidae) in the Miocene. This great difference is said to devalue various statements regarding the history of the mammals and man. "These considerations lead further to the taxonomic problem as to whether different groups of the same formal rank, for instance families, are or can be made equivalent by some criterion such as time of origin." He does not discuss this in detail, "but it may be affirmed that this time criterion, at least, is impracticable and quickly leads to confusion and absurdity." "Absolute equivalence between families (or other units) of different zoological divisions, such as fishes and mammals or, *a fortiori*, insects and mammals, does not exist and is probably quite unattainable."

It is very regrettable that Simpson does not explain why use of the time criterion is "impracticable" and leads to "confusion and absurdity." So far as I can see, this impression is based primarily on three reasons.

The first reason is probably the tacit assumption that requiring that the absolute rank of systematic groups be determined by their age imposes demands that cannot be fulfilled—absolutely equal age of origin would have to be proved before certain subgroups of different phyla could be designated as order, families, subfamilies, etc. This goal is, of course, scarcely attainable even in the distant future. We need only visualize that in a group of insects with 100,000 species there must be tens of thousands of intermediate stages between the category "order" and the species category if all kinship relations are to be represented truly and expressed in the system. It takes little imagination to perceive that it is probably never possible to determine which of these categories actually originated at the same time as certain categories of another insect group that is equally rich in species and equally differentiated. If this is what is meant, then fixing absolute rank according to age of origin is undoubtedly impossible.

It may be useful to remember, however, that historical geology is in a position similar to that of phylogenetic systematics: it faces the problem of relating sequences of formations on different continents. Here too this is done according to the principle of "simultaneity." Answers to many important questions depend on whether and to what extent the correlation is carried out successfully. Henbest (1952) pointed to the great difficulties faced by the historical geologist: "Wadia (1919) states that an outstanding difficulty in the study of the geology of India is the difficulty of correlating accurately the various systems and series of rocks with the different divisions of the European stratigraphical scale which is accepted as a standard for the world." Du Toit (1959) says that "nearly every important break in South Africa, paleontological as well as stratigraphic, happens to fall somewhere within one of the European systems, and each more or less well-marked unconformity between the systems in the Northern Hemisphere is usually represented in this country by continuous deposition."

No one would conclude from this that historical geology should abandon the attempt to relate the sequence of strata of different continents according to the principle of equal age. It is only the exactness with which this can be done and the reliability of the conclusions that are affected by the difficulties that have been mentioned. The same is true of phylogenetic systematics. At least at present it is in a much less favorable position than historical geology in determining the "simultaneity" of events (in its case, the origin of systematic groups). But this merely means that, for the time being at least, and in the majority of animal groups, phylogenetic systematics must be content with a much coarser time scale for its correlations. The question is not whether certain groups that are to be given the same absolute rank arose absolutely simultaneously, but only whether they all arose within a certain period of time. For this purpose the history of the earth must be divided into several successive segments, whose boundaries need not necessarily coincide with those of the geological time scale. Animal groups that in all probability arose during one such time segment would then have to be called classes, for example, those that arose during a certain younger time segment would be orders, and so on.

Fortunately a certain lack of uniformity in the evolutionary rate of most if not all animal groups seems to favor the discrimination of such time periods. "If we place together the great geological periods, there appears to be something intermittent in the evolution of the organic world, even though we know approximately the great lines of continuity. It is to be assumed that cosmically connected environmental changes were at work" (Haase-Bessell 1941). Newell (1952) reaches very similar conclusions for the invertebrates: "Rise and fall in apparent evolutionary activity is not at random. In a large proportion of the major groups, times of low evolutionary activity tend to coincide. The periods of relatively low average activity are followed by times of relatively high extinction rates. These times of wholesale extinction were followed by renewed radiation into vacant or uncrowded ecological niches."

According to Newell, periods of high extinction rates were the Devonian and especially the Permian. Succeeding periods of differentiation were the Mississippian (following the Devonian period of extinction), the Triassic-Jurassic (following the Permian period of extinction), and even before this the Cambrian-Ordovician. Mägdefrau (1942) describes a similar situation for plants: "If we enter the appearance and extinction of plant groups in a stratigraphic table, we find that there were times in which numerous new forms arose, that many families in fact differentiated explosively. This was the case at the turn of the Upper Devonian–Lower Carboniferous, and at the end of the Lower Cretaceous." When we recall that we regarded it as possible in most cases to decide whether a group arose before or after the end of the Lower Cretaceous, before or after the end of the Early Tertiary, then a subdivision of the earth's history like that shown in Fig. 58 may prove to be suitable.

The number of distinguishable category stages greatly exceeds the number of these distinguishable time periods (see for example the category tables of Simpson 1945, Kiriakoff 1948, Hargis 1956, all of which name only some of the most important categories). Consequently a whole group of category ranks must be assigned to each time period, and correlation according to age of origin would not be possible within these groups. This assignment is facilitated by the fact that it is customary to group the individual categories terminologically around certain main categories (the so-called obligatory categories). Consequently we would have to assign a certain category group (for example, the phylum, class, order, family, etc. stages in the terminology of Poche 1912) to each of the different time periods (I-VI in Fig. 58). This certainly cannot be achieved overnight, so it may be useful to show at least within a particular group how a certain clarification can be attained on this basis. Fig. 59 is a phylogenetic tree of the insects, taken with minor changes from Hennig (1953). I mention this because the relations shown scarcely differ from those shown in a similar phylogenetic tree proposed by Ross (1955). Here we are interested only in the data that may be important in determining the absolute rank of various groups. The oldest fossil insects come from the Middle Devonian. They belong to a monophyletic subgroup (Collembola) to which only a relatively small part of the modern insects belong. From this we can infer with certainty that in the Middle Devonian there were other subgroups of insects to which recent species also belong. There is no reason to assume that the insects themselves arose prior to the Cambrian. Since they are generally considered to be a "class," the time period II (Fig. 58) was assigned to the "class stage" for all the following. Therefore the Collembola, like the insects as a total group, would be one of the categories of the "class stage." The same applies to the Protura and Diplura, which likewise must have lived already in the Middle Devonian if the relationships indicated in the diagram are correct. Since the Collembola, Protura, and Diplura form a monophyletic group (the "Entognatha"), their sister group must have also existed in the Middle Devonian. This gives the following differentiation, within the "class stage," of the oldest subgroups of insects:

Class Insecta
    Subclass Entognatha
        Infraclass Ellipura
            Microclass Collembola
            Microclass Protura
        Infraclass Diplura
    Subclass Ectognatha

This classification is based on the following assumptions:
1. That all the groups are monophyletic.
2. That the kinship relations (sister-group relations) between these groups are as shown in Fig. 59.
3. That the fossils described from the Middle Devonian (*Rhyniella praecursor* Hirst & Maulik) are truly Collembola.
4. That the absolute rank of the groups has been determined according to their age, and that thereby the "class stage" is assigned to the time period from the Cambrian to the Devonian (II in Fig. 58).

The category hierarchy within the class stage remains arbitrary. This makes it possible for the time being to reserve the well-known category designations ("class" and "subclass") for the important and morphologically isolated groups. All possibilities of future refinement of the method remain open.

The prerequisites for subdividing the subclass Ectognatha are not so favorable. No fossils earlier than the Upper Carboniferous (time period III) are known. Consequently at present we can determine only which subgroups must receive

VI	Miocene
V	Oligocene
	Upper Cretaceous
IV	Lower Cretaceous
	Triassic
III	Permian
	Mississippian
II	Devonian
	Cambrian
I	Precambrian

Figure 58. Division of earth history for the determination of the absolute rank of the systematic categories of higher order. For explanation see text.

rank designations from the ordinal level, but not whether any must be designated infraclasses or microclasses as in the Ectognatha. But even this clarifies things somewhat. For example, no subgroup of the Holometabola is known from an older time than the Lower Permian, and it is therefore very questionable whether the stem line of even one of the subgroups of the Holometabola customarily called orders goes back to the Upper Carboniferous. This is even questionable for the Holometabola as a whole. On the other hand, the two subgroups of the Saltatoria (Caelifera and Ensifera) appear to have been identified as far back as the Upper Carboniferous. If we continue giving ordinal rank to those subgroups of the Holometabola that are customarily called orders, this would mean that the Caelifera and Encifera would have to be given at least the rank of "order"—as many authors in fact do—and that we could not follow authors who maintain that the Saltatoria (Caelifera and Ensifera) or even the "Orthoptera" (Notoptera + Phasmida + Saltatoria) represent a single order. But this decision actually goes beyond the principle of assigning at first only certain category groups to a particular portion of time, and already indicates a certain refinement of this method. Consequently the following example is even more convincing.

Many authors distinguish three "orders" of "Psocodea": the Psocoptera, Mallophaga, and Anoplura. Others lump the Mallophaga and Anoplura in one order, the Phthiraptera. It can be shown that the "Phthiraptera" are very probably a monophyletic group that probably did not arise before the Mesozoic (Königsmann 1960). The total group Psocodea can already be identified in the Lower Permian, but all that can be said of these fossils ("Permopsocida") is that some of them could very well be the ancestors of all recent "Psocodea." What we said above (p. 168) concerning the limited value of relatively plesiomorphous fossils applies to them. At any rate there is not the slightest reason for assuming that the "Permopsocida" are closely related to any of the recent subgroups of the Psocodea. Consequently only the Psocodea as a total group can come from the time period to which the category group "ordinal stage" was assigned, and only the Psocodea (and not the Psocoptera and Phthiraptera, or even the Socoptera, Mallophaga, and Anoplura) can be designated an "order." In this example there is also serious doubt that the Psocoptera and Mallophaga are monophyletic groups, the characters being based on symplesiomorphy.

The preceding examples seem to show that fixing the absolute rank of monophyletic groups on the basis of their age would not lead to radical changes in currently accepted rankings, but only to a safely founded decision in individual questions where differing opinions have been voiced by different authors. But the picture changes if we use other groups for comparison, even within the so-called insect "orders." It is very probable, for example, that the approximately fifty "families" of the Cyclorrhapha (Diptera), or at any rate most of them, did not arise prior to the Upper Cretaceous (time period V in Fig. 58). On the other hand we know for sure that many "families," and even groups that many authors consider subfamilies (e.g., among the Fungivoroidea), already existed in the Lower Jurassic (Lias) (see Hennig 1954). It is interesting that Handlirsch (who was anything but a follower of consistent phylogentic systematics) already sensed that the families of the "Cyclorrhapha" and the Nematocera were not equivalent and united a large part of the cyclorrhapid families (the so-called Acalyptratae) into a single family. He has scarcely been followed in this, although actually he was right (if we overlook the fact that he also proceeded in the opposite way and dissolved the families of the Nematocera and lower Brachycera and thus could adapt them to those of the Cyclorrapha).

All the so-called "orders" of placental mammals arose in the Upper Cretaceous (see the phylogenetic tree in Romer 1950). Thus in age they are comparable to the families of the Cyclorrhapha and not to the orders of insects. If absolute rank is based on age, and the time period between the Upper Carboniferous and the Upper Permian (III in Fig. 58) is regarded as the "ordinal" stage as was done for insects, and the period between the Triassic and Upper Cretaceous (IV in Fig. 58) as the family stage, then the mammals (more accurately the Theromorpha of von Huene) would have to be called an order (like the orders of insects) instead of a class. The Marsupialia and Placentalia would have to be downgraded to families, and the "orders" of the Placentalia would be tribes if the time span between Upper Cretaceous and Oligocene (V in Fig. 58) is assigned to this category stage.

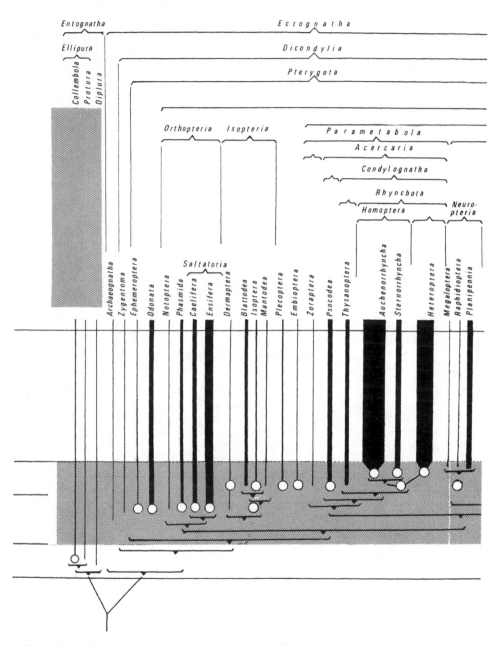

Figure 59. Phylogenetic tree of the insects. In this draft all those monophyletic groups are entered whose origin in the paleozoic has been demonstrated or may be assumed probable. By shading, the time interval (Carboniferous to Permian) has been emphasized; it corresponds to the section III of the earth history as depicted in Fig. 58. By means of shaded stripes, upper left and right, it has been indicated that the origin of all constituent groups of the Ectognatha (right), as far as presently determinable, falls into

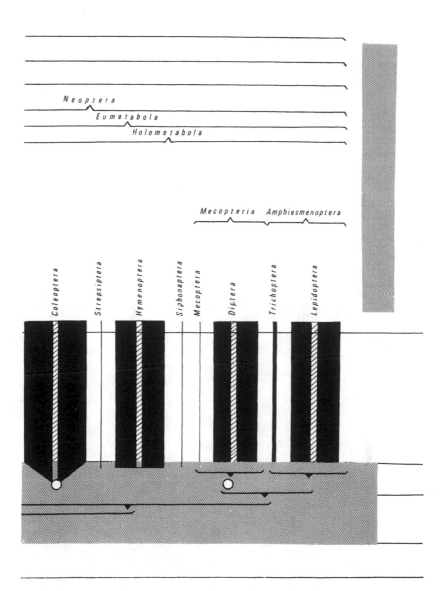

this time interval; on the other hand the constituent groups of the Entognatha (left) and the whole group Ectognatha are older. The approximate numbers of species of the related groups considered are indicated by the width of the lines and stripes. This does not apply for the four holometabolic groups Coleoptera, Hymenoptera, Diptera, and Lepidoptera. Each of these includes 100,000 or more species. For further explanations see text.

The re-evaluation would be even more drastic for some other groups. Certain of the so-called genera of the Ostracoda (Crustacea: *Bythocypris, Bairdia, Macrocypris, Pontocypris, Cytherella*), for example, apparently are known with certainty as far back as the Ordovician. This means that each of these "genera" would have to receive a higher rank (one from the "class stage") than the entire group of land vertebrates (Tetrapoda), which probably arose in the Upper Devonian.

This is probably the second reason why determination of the absolute rank of systematic categories on the basis of their age seems impractical and absurd to most zoologists. But we should take the trouble to consider cooly and without prejudice what is actually so absurd if several classes of ostracods are distinguished but all the land vertebrates are regarded as representing a single class. The groups would still remain if they are truly monophyletic; only their rank designation would be changed. We discussed above what would be gained. What would be given up is a last typological component of the phylogenetic system. Looked at carefully, the "absurdity" consists only in the fact that, in changing the rank of systematic groups to correspond with their age of origin, a principle would finally be given up that basically was already given up when we decided in favor of the phylogenetic system instead of the typological or morphological system. In the typological system the (morphological) success of phylogenetic development determines the formation of the groups themselves— and consequently the construction of the system itself—long before the question of the absolute ranking of the categories comes up. Species that appear to represent one and the same "type" are grouped together, in the last analysis without any regard to their kinship. To such a systematics the phylogenetic system must seem absurd wherever the boundaries of monophyletic groups cut across the group formations of the typological system. In this sense the phylogenetic system is absurd to the typologist when it unites the crocodiles and birds as sister groups in a group "Archosauromorpha" (von Huene), because the boundaries of these groups cut straight across the typological group formation "Reptilia" (which includes the crocodiles). The obvious morphological types or "structural plans" by themselves mean little in the phylogenetic system, whereas the segments of the phylogenetic tree mean everything. The phylogenetic system does not regard "equivalent" morphological types, but segments of the phylogenetic tree of equal age, as comparable units and tries to express this clearly and unmistakably in the relative and absolute rank orders of its groups. To typological thinking all this must seem abstract and unclear, all in all "absurd." It makes little sense, however, to follow the argument between typological and phylogenetic thinking, between typological and phylogenetic systematics, through all individual questions. The decision between these two systems is based on other considerations. Anyone who has decided in favor of the phylogenetic system should not reject its principles in individual respects, and if he does so he will be refuted merely by the inconsequence of his action. This will be true even if the "absurdity" of the phylogenetic system becomes evident to him only in the matter of absolute rank order.

The requirement that rank designations must express the comparability of categories—however remotely related the groups—is not a fundamental principle of phylogenetic systematics to the same degree as the requirement that the system must contain only monophyletic groups and that sister groups must be coordinate and be given the same rank. This fact makes a compromise possible, at least in a preliminary way. The idea of such a compromise is perhaps particularly advisable because taxonomists are essentially specialists—entomologists, arachnologists, ornithologists, etc.—who furthermore usually work only in certain sections of these extensive areas. All these specialists work as if only their group of animals existed. Consequently each specialist can erect a consistent phylogenetic system for his group without any necessity for correspondence on the basis of equivalent age between the absolute rank order of his categories and the absolute rank order of other groups of animals. Presumably even the most convincingly presented objective reasons will not bring these specialists to the point of giving up life-long habits and speaking of classes and orders where they are accustomed to speaking of families and vice versa.

The suggested compromise is to designate different time scales for different animal groups that would make it possible to retain the present absolute ranking of most subgroups. For example, for mammals (and probably also for birds) the time portion V (Fig. 58) could be assigned to the "ordinal stage" and the time portion IV to the "class stage." The present rank hierarchy of groups of mammals would scarcely be changed. For insects, on the other hand, the time portion III could be assigned to the "ordinal stage" and the time portion II to the "class stage." With the help of a conversion chart we could determine that the "orders" of mammals and birds cannot be compared with the "orders" of insects, but perhaps with the "tribe stage."

Such a compromise would certainly have all the disadvantages of any compromise, but it would be a true compromise not overdemanding to the specialist but still creating a basis for comparing "equivalent categories" in different animal groups. It would also leave open all possibilities for future improvement of the comparative basis. At any rate it would be preferable to the condition described by Simpson: "Equivalence between families (or other units) of zoological divisions, such as fishes and mammals, or, *a fortiori*, insects and mammals, does not exist."

The third reason that can lead to the impression that fixing absolute rank on the basis of the age of the group is "impracticable" and would soon lead to absurdity is probably the consequences that would result from strict application of this principle to fossil groups. If, for example, we assign with universal validity the "class stage" to time portion II (Fig. 58) and the "ordinal stage" to time portion III, this would mean that in a group that became extinct at the end of the Permian no "families" or other lower categories could be distinguished. Naturally a hierarchic system with coordinated and subordinated subgroups based on the relationships of the species could be erected for such a group, as for any recent group. But each terminal twig in the phylogenetic tree of the fossil group would be not only a species, but at the same time a representative of a higher category, such as the "ordinal stage." This would be absolutely correct and meaningful in a system that includes both fossil and recent forms, because if each fossil species had modern descendants they would be representatives of separate orders (or other categories from the "ordinal" stage). The fact that the fossil species occurs as the sole representative of a high category means that it became extinct without further splitting up. This is "absurd" only insofar as it contradicts ideas associated with our more or less typological way of thinking of the higher categories. Formal nomenclatural reservations could also be advanced here, but these only seemingly exist and can easily be removed. In order to meet existing habits of thinking, a compromise similar to the one proposed above for correlating the absolute rankings of different animal groups is possible here. I would not like to go into this any farther to avoid going too far afield. The following applies to all objections to the phylogenetic system that paleontology has raised in this and other respects:

The basis and point of departure for systematics is the modern animal world. Why we need a phylogenetic system of recent species has been thoroughly discussed, as were the structural principles of the phylogenetic system. Insofar as fossils can help erect such a system, they are welcome. There can be no objection to the classification of fossils in the system, since fossil and recent species are portions of one and the same phylogenetic tree, but the inclusion of fossils must not violate the structural principles of the system. There are, however, countless fossils whose inclusion in the system produces difficulties, particularly when incomplete preservation and other inadequacies with which paleontology must contend makes recognition of exact relationships impossible. In these cases the fossils may be more of a burden than a help to phylogenetic systematics. In no case should consideration for paleontology lead to abandonment of the principles of phylogenetic systematics.

It is noteworthy that opposition to the phylogenetic system often results from the problem of classifying fossils—although the difficulties are only formal or technical—whereas there is never any opposition to the properly understood theory of phylogenetic systematics. The difficulties that fossils present to the theory of typological systematics—insofar as such a theory exists at all—are overlooked. Suppose, for example, that numerous fossils were known from the time period during which the birds separated from their sister group (which in the typological system is placed among the reptiles). If our ideas of the course of phylogeny are even generally correct, the differences between some of them would have to be as minute as the differences between modern species that only recently split. Nevertheless one of these barely different fossil species would belong to the bird line, the other to the sister group of the birds (typologically placed among the reptiles). It is not evident where typological systematics would draw the line between the reptiles and the "bird type" without either deciding arbitrarily with which character expression the "bird type" begins or utilizing the principles of phylogenetic systematics. In dealing with the classification of fossils, typological systematics lives on the incompleteness of the fossil record. The more numerous and completely preserved the fossils, the greater the difficulties of typological systematics become, whereas the difficulties of phylogenetic systematics decrease. This should never be forgotten in discussing the classification of fossils in the phylogenetic system.

THE ARRANGEMENT OF TAXONOMIC GROUPS IN THE SYSTEM

Taxonomic work results in a hierarchic system of groups. The content of the groups is determined completely unequivocally by the facts that they *must* include only living species descended from a common parent species, and that they *must* include all living species descended from a common parent species. The ranking of all these descendant groups is determined by their age as independent groups. Consequently the relationships expressed in the group structure of the phylogenetic system are linear and presentable in one dimension. The representation of sister-group relationships is decisive in the phylogenetic system. Because of their equal age, sister groups are of equal rank: they are coordinate and completely equivalent. Naturally, however, they cannot occupy the same spot in a presentation of the system; they must be treated one after the other, or drawn side by side. This raises the question (to be sure very subordinate and formal) whether there is a rule for deciding which of two or more sister groups should be treated first. With regard to the already repeatedly mentioned rule of deviation, of two coordinate groups we would deal first with the one that remained most similar to the parent species, and which for this reason we have designated "plesiomorphous." But of two sister groups, one is not plesiomorphous wholly and in all characters while the other is apomorphous. The argumentation scheme of phylogenetic systematics (p. 90) shows that each monophyletic group must have apomorphous characters in order to be recognizable as monophyletic, and consequently each of two well-founded sister groups is likely to have at least one apomorphous character. Consequently it is not always possible to decide which group is more plesiomorphous. In addition there are all those cases of groups that include several subgroups which at least at present appear to be equivalent because it is impossible to decide whether there is a closer relationship between some of them than between others. But even when it is clearly evident which of two sister groups is more strongly plesiomorphous and which more strongly apomorphous, some species in each group resemble the parent species more closely (plesiomorphous) than do other ("derived," apomorphous) species of their own group. This is because the rule of deviation has been active at all times, and consequently its operation is reflected at all category stages of the hierarchic system. If two coordinate groups are dealt with one after the other— which is unavoidable—then naturally the plesiomorphous species of the second (on the whole more apomorphous) group must be placed after the relatively apomorphous species of the first (on the whole more plesiomorphous) group. This circumstance has been criticized as a defect of the phylogenetic system, or even as "proof" of the impossibility of a logical phylogenetic system, but naturally this is completely incorrect. We could make the same objection to any verbal description of a planar or spatial structure, where it is likewise necessary to progress from a point first in one and then in other directions without regard to the fact that neighboring points often become widely separated in the description. Description or any other verbal presentation of the phylogenetic system must be regarded as unobjectionable if it expresses correctly the coordinate and subordinate relations of groups according to the principles discussed in the preceding sections. The sequence of coordinate groups is without significance, and therefore need not correspond to the average "level of organization" of its members. All relationships between coordinate groups can be clearly expressed in the written description; they need not be apparent in the arrangement of the groups. Hence it makes no sense to characterize systems as different—as is actually often done—merely because the same groups of systems are arranged a little differently. This is probably a lingering effect of the ancient concept of the "ladder of organisms," expecting that a developmental series from "lower" to higher, or at least from more "primitive" to derived forms, must be expressed in the phylogenetic system. But the task of the phylogenetic system is not to present the result of evolution, but only to present the phylogenetic relationships of species and species groups on the basis of the temporal sequence of origin of sister groups. Our investigation of the morphological methods of phylogenetic systematics, and the "scheme of argumentation" (Fig. 22) developed from it, also showed that there can be no groups with either exclusively plesiomorphous or apomorphous characters, and that consequently "specialization crossings" * are a prerequisite for recognizing monophyletic sister groups as such. The amazement still expressed by some at the existence of such "specialization crossings" is likewise explainable from the old prejudiced notion that evolution had to produce a simple ladder of organisms.

It is often assumed that presentation of relationships in the graphic form of a phylogenetic tree is superior to a written fixation of the system in the form of a catalog or monograph. The phylogenetic tree may make things more obvious, but any fact or supposition that can be entered in a tree can also be expressed unequivocally in a "written fixation of the system." The fact that two or even three dimensions can be used in presenting a family tree is only of slight importance, and is true only in the limited sense that the "results of evolution" can be expressed in the additional dimensions, in addition to the "phylogenetic relationships." We can probably agree in principle with Zimmermann's statement that the family tree, which has had "a curious fate that swung back and forth between worship and scorn," is just as indispensable as a means of communication as is the curve in physiology. But it should be emphasized that the curve

---

* Spezialisationskreuzungen. See for example, O. Abel, Palaeobiologie, 1912, p. 639.

Figure 60. Presentation of phylogenetic tree that does not permit ready recognition of to what extent sister-group relationships have been expressed.

in physiology is a graphic presentation of facts that can be expressed differently (e.g., as an equation). Our contention is generally true, that "family tree" and "written fixation of the system" correspond exactly only when the family tree clearly shows recognized or presumed sister-group relationships, and also makes clear which groups are undoubtedly monophyletic and which are doubtfully so (see Fig. 47, for example). By no means all of the family trees scattered through the literature satisfy these criteria.

In the family tree shown in Fig. 60, for example, it is not evident whether we can assume sister-group relations between the genera *Amblyomma* and *Alloceraea* + *Aponomma* on the one hand, and then between these and *Haemaphysalis* on the other; or whether *Amblyomma* is a grouping based on symplesiomorphy from which the other genera arose by "radiation." Furthermore the typological grouping of the Ixodidea at the head of the figure does not correspond to the family tree presentation.

Fig. 61 is also ambiguous. We cannot tell whether the way in which the group Tetrarhynchidea is entered is intended to mean that it is a group based on symplesiomorphy and that the monophyletic group Pseudophyllidea arose from the Tetrarhynchidea by "radiation," or whether a sister-group relationship between the Tetrarhynchidea and Pseudophyllidea is assumed. The way in which the Tetrarhynchidea are entered may only be intended to express that this group has a great many plesiomorphous characters relative to the Pseudophyllidea, so that we can assume that the ancestors of the Pseudophyllidea looked in general much like the recent Tetrarhynchidea.

Both forms of presentation are widely used, but they should be avoided in the interest of clarity.

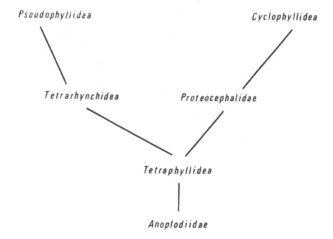

Figure 61. Presentation of phylogenetic tree that does not permit ready recognition of to what extent sister-group relationships have been expressed.

III

Problems, tasks, and methods of phylogenetics

GENERAL

The Concepts of Evolution and Phylogenesis

Since the acceptance and conclusive foundation of the idea of evolution, investigation of the course of evolution and its possible laws has become one of the most important tasks of biology. Evolution is change. When we say an existing condition arose by evolution we mean that it originated from a different previous condition. The smallest biological units that undergo change are the individuals. "Change in the manifestation of the individual" (Prell 1929) is consequently the basic process of biological evolution. Systematics is not participating directly in its investigation. The evolution of organisms receives its very special character —which is also decisive in the tasks of systematics—from the existence of species, and the evolution of organisms thereby becomes phylogenesis. The concepts of evolution and phylogenesis are rarely sharply distinguished. It is to be emphasized that the term phylogenesis can refer only to the totality of change in things, i.e., of individualized natural bodies, and even in general only insofar as this change is connected with a splitting or division of these individuals. Consequently we can also use this concept where, from an individual in the narrow sense (i.e., a single organism), daughter individuals and from these granddaughter individuals, etc. are formed without sexual processes necessarily taking part in species differentiation. In this case the "stock" to which this phylogenesis leads is the totality of the individuals descended from an original individual. Normally, however, the evolution of living nature is by way of species. As we have shown, species have a character similar to that of individuals. Only within the frame of species differentiation do the individuals achieve reproduction. The division of species is the process by which self-differentiation of living nature takes place. The common character of the organisms belonging to a stock or "line"—i.e., to a community of descent—is that they originated from a common stem *species*. Thus the concept of phylogenesis is intrinsically tied to the concept of species splitting and the consequences of this process. Consequently, strictly speaking we can speak of the evolution of an individual character—e.g., the wing venation of insects—but not of its phylogenesis, since a character has no actual individual nature but appears only as a peculiarity of an individual. Thus the evolution of a character as a concrete process is tied to the phylogenesis of the organisms, but since the same character can develop in many different phylogenies (see p. 117) it can also be followed on the basis of the purely formal sequence of its stages of development, even though the bearers of the various evolutionary stages are not closely related phylogenetically (stage series in contrast to ancestral series).

The unsharp distinction between evolution and phylogenesis, or failure to observe the fact that the evolution of organisms is exclusively by way of phylogenesis although "evolution" is not identical with "phylogenesis," seems to me an important reason for the misunderstanding and negative attitude often found even today among representatives of the humanities and related disciplines. Apparently the idea of a somewhat undisciplined development in which anything can happen, or in which so to speak there is a constant transition from "lower" to "higher" on a broad front, is often associated with the concept of evolution. Even Mühlmann (1939) does not seem to have been free of this idea and refused "biological evolution" (or what he considers this to be) for human ethnology, or at any rate assigns it a subordinate position. In this sense he and others contrast evolution with historical events. They overlook the fact that evolution as *phylogenesis* is an eminently historical process that follows absolutely fixed laws, but one that takes place in individualized beings and consequently in any given case has—within the framework of general conformity to law—the character of the nonrepeatable individual. I believe this fact should be more clearly emphasized in view of the misunderstandings to which the concept of evolution still seems to be subjected by outsiders.

Our assertion that the basic evolutionary event is change of form in an individual might suggest that investigation of the general laws of evolution could be limited to the modifications of organisms, and perhaps the altering of species structure and species cleavage resulting from such individual modifications. Genetics, especially evolutionary genetics, would then be the only biological science that could deal with phylogenesis from the nomothetic point of view. Such an idea actually seems to be widespread, particularly among geneticists. Thus, Timoféeff-Ressovsky says: "In spite of the fact that in every case it starts inductively from described material, the classical period of evolutionary study must be regarded as predominantly deductive. It is interested primarily in large groups of organisms, long periods of time, and higher systematic categories. The second period, by contrast, starts from the most rigorous analysis—microevolution—and ascribes decisive importance to the small and smallest evolutionary steps" (cited in Haase-Bessell 1941). Elsewhere the same author (1939) says bluntly that since "macroevolutionary investigation" has exhausted its methods, microevolutionary investigation can now take over with new methods and fresh forces."

We agree with these views only to the extent that phylogenesis actually can be understood only if "microevolution," the origin of small and smallest evolutionary steps at the individual level, is known. But we must emphasize strongly that this does not provide a complete understanding of the evolutionary process and its laws. It is generally recognized today that the processes of form change in individuals or the mutation process certainly does not explain phylogenesis completely. The fate of a mutant in a breeding population determines the role it will play in phylogenesis, but this is not the end of the process. Evolution does not become phylogeny through a single process of species splitting. Phylogenesis is the origin of groups of species from a stem species and its descendants by progressive splitting, resulting in a later evolutionary condition than that of the original stem species. This raises the question whether comparison of as many such monophyletic groups as possible, of the same and different ages, would not reveal conformities to law beyond "alteration of the form appearance of the individual" and the individual cleavage process.

We may now ask whether the conformities to law revealed by investigating the evolutionary process of higher taxa can be compared with the "laws" of the other natural sciences, or whether they differ in some essential respect. Views are very divided on this. It is often believed that phylogenetic investigation, at least of higher taxa, cannot reveal causal relations, but only functional relations. If this were true, phylogenetics could not be called one of the sciences working from the nomothetic point of view. At first sight such a suspicion appears to be well founded if we start out from the presentation of phylogenetic work that was given above. Comparing—actually or only in principle—the degree of phylogenetic relationship with the presence of other relationships (morphological similarity, vicariance, etc.) actually seems capable of determining only functional, correlative relations, not causal relations. But this concept is only partly correct. If such a comparison reveals a correlation between vicariance and degree of phylogenetic relationship this is primarily only functional or correlative, but if it is found further that the most closely related units usually vicariate whereas more distantly related units do not then a causal relationship is indicated—we may infer a causal relationship between the differentiation of space and the origin of differentiation in one of the groups occupying this space. Further investigation of this relationship is likewise a phylogenetic problem, but how do we go about it? We think first of the methods of genetics and population research, and are partly justified in this since we have emphasized that all evolution begins as a change in the form of individuals and that the fate of these changes is decided in the action field of the species assemblage. Consequently a knowledge of the causal processes involved in the form change of individuals and in changes in the fabric of the species belongs to a knowledge of the causal processes—the mechanism—of evolution. But it is a mistake to assume that this is enough. Large scale evolution is the result of the interaction of many and varied causal processes that participate in various ways in altering the form of species. We can no more understand the total process from a knowledge of the individual processes than we can learn the laws of growth of a forest from a knowledge of the growth processes and growth conditions of the individual plants. It is often emphasized that understanding an event resulting from the interplay of many individual processes requires knowledge of the integral laws of interaction in addition to knowledge of the individual processes. Von Bertalanffy (1932), in particular, often noted this. From similar considerations Morgan (cited from von Bertalanffy) distinguished between emergent and resultant evolution: "In emergent evolution every stage (atom, molecule, colloidal unit, biokyl, cell, multicellular organism, society of organisms) has characters that cannot be derived from those of the subordinate elements—in contrast to the mere resultant." Of course these considerations apply in biology primarily to "totalities," "forms," and "stages" as represented by individual organisms, and to societies and similar "supra-individual entities," and less so to higher taxonomic categories. But the concepts totality, forms, system, and stage—which play a role in all these considerations—have been extended to biocenoses and other ecological systems (e.g., Friederichs 1930, Thienemann 1940). And the higher taxa have an individual-like character and are basically indistinguishable from species (see p. 83). These facts suggest that we should at least be cautious and prepared to find features in the evolutionary processes of higher taxa that cannot be calculated as simple resultants of the interaction of partial processes. Nothing is altered in this presentation if we follow the statements of Bünning, who opposed the idea that lies at the bottom of all the above arguments, namely that "the whole is more than the sum of its parts"—whether the "whole" is a structure (organism) or a process in which, as in phylogeny, numerous partial processes act together. Whichever view we accept, for our purposes these differences of opinion are only different formulations that—however important their distinction for other purposes—agree in stating that a process representing the interaction of several partial processes cannot always be fully understood on the basis of a knowledge of the partial processes alone.

Therefore it is certain that knowledge of the elementary processes alone is not enough for investigating the grand strategy of evolution. This need not be because we do not know the integral laws of their interactions in the sense discussed above. The individual processes involved in an evolutionary process are often so numerous and varied that we may be able to grasp them in principle and in a few model examples, but we cannot do so in any given concrete case. And aside from the practical impossibility of surveying them in detail, it is not only a knowledge of all individual processes that interests us, but also the questions of which of them play the major role in evolutionary events, and which of their many possible interactions can be regarded as typical. This information cannot be derived deductively from a knowledge of the separate processes. We know, for example (see the formulas of Wright), that isolation necessarily leads to differentiation of the severed populations of a species, and thus ultimately leads to species cleavage. We also know the mechanism—i.e., the elementary causal processes and their typical interactions—on which this process rests. It is clear too that spatial isolation of parts of a group of organisms would also isolate any parasites living on them, but it would be completely false to conclude that parallelism in the differentiation of host and parasite groups would necessarily follow. The question of whether such a law of evolution is actually valid, as has been asserted ("Fahrenholz' rule," according to Eichler), requires special investigation. A particularly beautiful proof that the laws of evolution cannot be derived deductively from the action of separate evolutionary factors is Willis' "age and area" theory, according to which the area of a group and its species number are proportional to the age of the group. It is evident that the number of species in a group must increase, within certain limits, in the course of its evolution (see p. 182). We can demonstrate the mechanism that brings this about (the individual causal processes) without difficulty in model examples. It is also certain that all species "tend" to extend their area, and the causal factors for this are also known. They are doubtless important causal factors, together with the factors leading to species cleavage in the course of time. Even if we knew them completely, however, we could not derive from them a law such as Willis' law by deduction. Determining the significance of individual factors in the total evolutionary process is the unique task of phylogenetics, and this raises two important questions:

1. What methods are available to phylogenetics for solving this problem, and what perceptive value do they have?
2. Are all the laws that have been derived—all of which have the character of Willis' assumption and the "Fahrenholz rule"—to be designated causal laws (aside from the correctness of these two special laws), and what is their position in the general system of natural law?

These two questions are internally connected. With regard to the first, it is clear that we are not talking about experimental methods. At best, only processes in the action field of the species category can be influenced experimentally. The scope and temporal span of the evolutionary processes involved in our problem are so far removed from human influence that they will never be solved by experimental methods, or even by direct observation. Apparently only the method that Max Hartmann (1927, 1932) called the "comparative method," as sharply opposed to the experimental method, remains. According to Hartmann (1932) both methods have "their logical bases in the two different kinds of induction that apply quite generally in all natural sciences: pure or generalizing induction, and exact induction." Of the capabilities of the two methods Hartmann says that "the comparative method primarily reveals only similarities and differences in various biological objects, phenomena, or processes" and "tries, on the basis of subsequent causal or teleological connections among the concepts so obtained, to find not only order but also conformity to law. As a rule it is not capable of rigorously proving the latter, and consequently leads mainly to characterizing of phenomena and processes and correct posing of the problem. By contrast, exact induction—the truly inductive or analytical method for discovering lawful connections between phenomena or processes—proves to be the more compelling, more penetrating, and therefore more fruitful and inventive method."

But is it really fair to contrast the comparative and experimental methods so sharply, or to say that phylogenetics is restricted to the comparative method, at least in dealing with the broader evolutionary processes? The core of this problem and its solution seem to me to be found in the statements of Mühlmann (1939). He starts from the premise that the insights of modern physics, "reduced to the most general formula," show "that physical laws have a statistical character, that is they are statements of probability. If this be true, then the difference between experimentation and statistics disappears first of all in the area of the theory of perception. The experiment, which hitherto has been the strongly favored method of natural science, no longer differs basically from the experiment 'made' by an ethnologist when he studies the sequence of situations involved in the collision between a Papuan community and European civilization. Expressed the other way around, the experimental attack is historical in its nature." If we transfer these views to the methodology of systematics we can go farther: every experiment starts from a suspected causal relationship. The suspicion usually originates through the use of the comparative method: it is assumed that one and the same causal sequence underlies a series of events between which similarities were recognized with the aid of the comparative method. In individual cases this causal sequence may be slightly modified by interference of other factors. Thus a hypothesis concerning this basic process is erected. The experiment serves to verify this hypothesis: the condition that is presumed to be the "cause" of another condition is artificially created, and the correctness of the hypothesis is considered proved if by creating identical starting conditions the resultant condition can be obtained at will.

We may now ask whether it is absolutely necessary that the starting condition be produced by the experimenter himself. No doubt this is necessary if we are seeking situations that are identical with the starting (causal) condition, where it makes no difference how they arose. If the expected resultant condition (the effect of the cause) appears, then in this case too the hypothesis must be considered as verified. But this kind of verification is also available for the hypotheses of phylogenetics. Suppose we have formulated, with the aid of the comparative method, the hypothesis that "with obligatory parasites the relationships among the hosts can usually be inferred directly from the systematics of the parasites" (Fahrenholz' rule), or expressed more appropriately to our context, that changes in the systematic differentiation of a host group bring on identical changes in the differentiation of the parasite group. It is not possible to test this hypothesis experimentally in the true sense. But it is possible to look for all cases in which a parasitized animal group has in the course of time undergone changes in its systematic differentiation. If we find that the parasite group is as we expected in all these cases we can consider the hypothesis to be as well verified as if we had carried out an experiment. The fundamental agreement between these two kinds of verification is even more evident when we consider that even in the experiment everything depends on creating truly identical starting (causal) conditions. Most discussions of the varied outcomes of allegedly the same experiment deal with the question of whether the starting conditions actually were identical and how difficult it is to satisfy these prerequisites. The difference in exactness between the actual experiment and one "conducted" by a phylogenist when he searches for identical initial conditions should not be argued away. Unquestionably it is vastly more difficult to demonstrate identical initial conditions than it is to create them for an experiment, but what is important here is that the phylogenist and the experimenter by no means use such different methods as might appear at first glance. In practice the "comparative method" is obviously not a uniform process either, but consists so to speak of two parts. The first is a truly statistical comparison of the phenomena, processes, etc. This comparison creates a pre-orientation leading to a hypothesis regarding suspected relationships (causal, but possibly merely correlative). The hypothesis is verified by experiments or by further comparisons, in which certain situations are now consciously searched for and compared in their sequences with those that are theoretically expected. There are practical differences, but none from the standpoint of the theory of perception, between the experiment and this second step in using the comparative method. The practical difference is important enough: an experiment can be repeated at will. Theoretically a one-time experiment is sufficient if the required initial conditions have truly been met; repetition merely serves to certify this prerequisite, and to demonstrate the accidental and unsystematic nature of small unavoidable deviations. The comparative method has this means of certification only to a very limited degree. In this case, too, it would be sufficient to find one case of the typical structure of the suspected basic process, but the comparative method can accomplish much less with this than the experimenter, who to a considerable extent has the creation of this condition itself under his control. In view of the individuality of all real phenomena, and therefore of all processes in the real world, the comparative method will not find two cases that actually are completely the same. The question is how far this is a limiting factor in the tasks of phylogenetics.

This brings us into the domain of the second question that we raised: are the conformities to law found by phylogenetics so exactly comparable to those of the other natural sciences that they can be called *causal* laws? It is often said that the conformities to law of phylogenetics—and this includes most of the so-called biological rules—must be contrasted as mere "rules" with the "laws" of other natural sciences, such as physics and probably also certain biological disciplines. In order to decide this matter it is necessary to know the definition of the two concepts. Horn (1929a) gives a simple explanation that agrees with customary practice: rules permit exceptions, laws do not. In a more detailed investigation Roux (1920) examined the distinction between "natural law" and "rule." He reduces the difference to the concepts of event (action) and occurrence. With regard to the event "the same combination of factors gives absolutely the same result in all places and at all times. Consequently the action is called 'conformity to law' only by analogy. On the other hand, when we speak of the occurrence of events we can do so only with respect to their temporal-spatial position and can speak only of the regularity of their occurrence." Consequently Roux believed that "in the future natural scientists can speak with complete exactness of laws of objective and temporal-spatial 'action,' and of rules and irregularities of temporal-spatial (partly also of objective) occurrences—in short, of laws of action and rules and irregularities of occurrences, but never of rules and irregularities of action." Thus to Roux rule and law mean

fundamentally different things. Natural law is a causal concept, a rule is a purely descriptive concept. A rule is the expression of a frequency relationship of occurrence; it denotes the corresponding event in more than 50 per cent of the observed cases. Natural law denotes an action of given factors; and since all actions are constant—i. e., take place everywhere and at all times in the same way under the same conditions—every properly determined and properly formulated natural law must apply without execption or it is false. On the other hand, the frequency with which the application of a law occurs—i. e., how often the named factors occur alone, without others participating—is completely meaningless.

Von Bertalanffy (1932) partly opposes the ideas of Roux. "The science of law does not consist, as is often claimed (Roux), in an insight into the causal necessity of the event." He sees the proof of this in the statistical nature of all laws of nature that have been recognized by modern physics. "A law can just as well be statistical as dynamic. Statistical laws are no less 'laws' than are the dynamic laws of classical mechanics." "A rule is a generalization from the description of empirical relations; a law is a statement deductively derivable from theoretical assumptions." Thus to him too rules and laws are basically different concepts.

In order to decide this question it may be useful to start from the investigation of a concrete example. The law of gravity is a famous and often-cited example of a natural law. As a simulated example of a phylogenetic "rule" we may choose the following statement: partition of the area occupied by a species leads to the splitting up of the species into a corresponding number of daughter species. An important difference between these two types of natural law is that there is a direct connection between the original cause and the final effect in the case of the law of gravity, whereas in the phylogenetic rule the relationship is indirect. Partition of the range of a species releases a whole chain of processes terminating in the splitting of the species. Would this statement for this reason alone have to be called something less than a causal law, if it had universal validity? Undoubtedly not, since there is no basis for restricting the concept of causal law to cases in which there is a direct connection between initial cause and final effect. There are still other important differences between the two examples: the law of gravity has unrestricted validity, whereas there are exceptions to the phylogenetic rule. But the law of gravity is in fact rarely, if ever, purely realized in the actual world. This is because, besides the factor of gravitation—which is the only one considered in the law of gravity—in the fall of a body there are always other participating factors whose actions likewise conform to law. By what right, then, can we speak of *the* law of gravity? Because within a class of natural phenomena the process accounted for in the law of gravity can be considered the basic process in all special cases. Modification of the basic process by other factors and the processes set in motion by these are so insignificant that they actually can be considered mere modifications. "Free fall," with which the law of gravity deals, is the typical case within a class, around which the special cases in this class are more or less closely clustered. How does this compare in this connection with our phylogenetic rules? Here too we are dealing with classes of natural phenomena, such as the effect of the partitioning of an area on the groups of organisms living in this area. Corresponding to the incomparably greater number of factors that interact in any such process, compared to the events in the fall of a body, the individual cases group themselves much less closely around a typical case. With much less justification can we consider a particular causal chain as the fundamental factor at the bottom of all observed phenomena in a particular class. Naturally it is possible to declare a particular random case as typical. We could then regard such a case as cause and effect, and in the isolation of South America, for example, we could formulate the following: with isolation of a portion of the earth's surface like that of South America during the Tertiary, if conditions are similar the effect on animal groups that are in a condition similar to that of the Simiae at the time of the South American isolation will be similar to that observed in the Simiae. Such a "law" would be merely the inference derived from an assumption of general causality. If everything left undetermined in the above formulation could be replaced by specific statements of conditions—which in principle is certainly possible—we would have a law whose absolute validity would have to be assumed in all absolutely equal cases.

But it is also possible to make an assumption concerning the general initial conditions in the partitioning of an area, and to deduce from this the effect on a group of organisms living in the area (concerning whose conditions assumptions would likewise have to be made). We would then have a deduced law in the sense of von Bertalanffy, and its general validity could be assumed in all cases agreeing with the assumptions. Such a law would have the same kind of general validity as the law of gravity. The difference from the latter is only that the individual concrete cases cluster much less closely around the "typical" case defined in the "law." The precision of the validity of a conformity to law can be increased in concrete cases if the typical situation on which the law is based approximates a particular concrete case as closely as possible and then narrows the class boundaries of the area of validity as much as possible. If the limits of validity of a "law" are increased by extending the class limits, then a restriction of precision must be taken into account. Limiting cases are the "absolute law." At one extreme it applies to an infinite number of actual or potential cases in nature because these are completely identical with the "typical" case with respect to the factors participating in the event. At the other extreme it applies to the absolutely unique event in which the combination of participating factors diverges so much from the combination found in all other events that the class limits must be drawn so broadly that the resulting classification represents no gain in our understanding. In our opinion, all conformities to law in the natural sciences lie between these two poles, the so-called "laws" lying closer to one pole, the "rules" closer to the other. Corresponding to the endless complexity of the objects and processes that occur between these two poles—i.e., corresponding to the distinct individuality, and thus to the historical character of the phylogenetic process—the conformities to law that can be recognized in phylogenetics should be called rules rather than laws. We have shown, however, that there is no contradictory contrast between the two concepts, and that there is no reason for the other natural sciences— whose conformities to law have more the character of "laws"—to look with contempt on phylogenetics.

These reflections may be almost pointless in view of the fact that investigations of "macroevolution" and "transspecific evolution" have become very modern again. This revival of interest would not be possible if their justification and importance were not recognized. It might be asked, however, what they have to do with the theory of phylogenetic systematics. The answer can only be that phylogenetic systematics and investigations of the "course and conformities of law of transspecific evolution" are inseparably connected by the principle of reciprocal illumination. In the following we must restrict ourselves to showing, for some of the basic questions of phylogenetics, how answering them is closely connected to the problems of phylogenetic systematics.

Monophyly and Polyphyly

Whether taxonomic groups arose monophyletically or polyphyletically, and just what is meant by the terms monophyly and polyphyly, has been much discussed from time to time. Many authors have maintained that a higher taxon can be called monophyletic only if its descent from a single pair of parents can be assumed, whereas others would call any group monophyletic that arose from a group of equal systematic rank.

Handlirsch (1925) and Mez (1926) represent the latter opinion, while Rüsch-kamp (1927) considered the opposite point of view correct: "If the generic characters are very pronounced in all species of the genus, and if these characters separate this genus sharply from the closest systematic genera, this suggests that the species of the named genus arose monophyletically, i.e. from a single pair of which at least one of the parents obtained the present generic characters perhaps by mutation. If this is the situation, then the present genus represents a phylogenetic unit."

Systematics is in no way free to make the decision in this controversy. It is determined by our conception of the actual course of the evolutionary process. As shown above (and as is generally accepted), it is decisive for this conception that the device of bisexual reproduction makes the species the real unit. New species can arise only through the breakdown of individually existing species. Consequently if (as was done above, consistent with the facts there presented) we define phylogenetic relationships among species in any way through the relationships of these species to a common stem species, this also defines unequivocally the concept of monophyly: only groups of species that can ultimately be traced back to a common stem species can be called monophyletic. Consequently in phylogenetic systematics only groupings that are monophyletic in this sense have any justification. As explained in detail above, to this definition it must be added that not only must a monophyletic group contain species derived from a common stem species, but it must also include *all* species derived from this stem species. If the question of whether the groupings of phylogenetic systematics arose in part polyphyletically makes any sense at all, it is only because it is not always possible to decide whether a supposedly monophyletic group did not actually arise polyphyletically, since systematics is not always able to guarantee the monophyletic origin of the groups it proposes.

The whole first section of this book was devoted essentially to answering this question, and we need not repeat what was said there. Up to now we have started out from the premise that "monophyly" means the origin of a group of species from a common stem species. This says nothing about the idea that the stem species and its successors must be divided into two equal or nearly equal parts in each successive cleavage. The insights of modern population genetics seem to support the idea that "new species" usually arise by edge populations splitting off from an existing species and developing into separate species in isolation. This changes nothing in the definition of the concept of monophyly, but perhaps it explains the "deviation rule" (p. 59), which is derived from the similarity distribution within species groups and which says that when a species splits, one of the two daughter species tends to deviate more strongly than the other from the common stem species (or from the common original condition). Special complications would arise if new species could also arise to a noteworthy extent by hybridization between species. No definite case of this seems to be known in zoology.

In botany, however (according to Schwanitz 1940), it has "been possible to prove experimentally in a number of cases that species in nature arose by hybridization and subsequent doubling of the chromosomes (*Triticum vulgare*, *T. durum*, *Prunus domestica*, *Nicotiana tabacum*, *Brassica napus*, *Galeopsis tetrahit*, and others). It has been possible in our own day to follow the spontaneous origin and spread of a species (*Spartina townsendii*).

From the standpoint of phylogenetic systematics it would have to be said that higher taxa (genera, families, etc.) formed by the cleavage of species that themselves arose polyphyletically are not to be regarded as polyphyletic units, because the criterion for monophyly—origin from a single stem species—would also apply to these units. Only the stem species would have arisen polyphyletically. The question arises, however, whether it is admissable to speak of a polyphyletic origin of species in this sense. Is not the species concept that the species includes all individuals that together are capable of producing completely fertile offspring, and must we not then consider groups whose individuals can produce new species by hybridization as partial groups of one species? If we speak of the origin of species by species hybridization, are we not guilty of circular reasoning between premise and conclusion? It was shown above that the criterion of fertile crossability alone is not enough for characterizing the species concept. In many respects it turns out to be too narrow, in others too broad. It was also shown that not all sides of the species concept can be used in the same way for all purposes connected with the use of this concept, without having to speak of a nonuniformity of the concept. The establishment of higher taxa and their monophyly is no doubt a measuring operation. The phylogenetic kinship of the species is measured, and the yardstick is the time that elapsed between the separation of the various species. Here, as in all other cases, the upper limit of the accuracy of this measure is determined by the relationship between object and yardstick. In our case this means that the accuracy attainable depends on the measurability of the species cleavage process. Since the temporal duration of this process varies from case to case, and particularly because the beginning and end of such a process cannot be specified exactly, it follows that a determination of the monophyly of higher taxa must necessarily remain inexact within certain limits. The taxonomy of higher taxa can easily make allowance for this uncertainty, since in all cases in which a "polyphyletic origin of species" has been recognized, the species involved were so closely related that they could just as well be considered races of one species. At any rate all observations indicate that the difficulties resulting from the indeterminacy of the species concept are so insignificant in the taxonomy of the higher taxa—i.e., in the evolutionary laws so disclosed—that they can be disregarded.

In summary, the definitions of the concepts of monophyly and polyphyly are given by the mechanism of bisexual reproduction, and thus by the existence of speciation. Since the evolutionary process is connected with a splitting up of existing species, we define as monophyletic groups those taxonomic categories that arose by cleavage of a particular single species, its stem species. This stem species must have monophyletic groups in common with no other group in the system. Polyphyletic groups do not satisfy these conditions, and therefore have no place in the phylogenetic system. Consequently the question of whether taxonomic groups are monophyletic or polyphyletic makes sense in the phylogenetic system only to the extent that we can ask whether it is always *practically* possible to recognize monophyletic groups as such. The first (methodological) part of this book deals with that question. On the other hand we must consider a polyphyletic origin of species in the category stages below the species to be possible, through hybridization for example, but this does not touch the question of the monophyly of the higher taxa.

Everything said above about the monophyly of taxonomic groups applies primarily to middle-rank categories. The problem of the origin of the highest categories, such as phyla, has not been touched. Lam (1926) would like to limit the term polyphyly (or biphyly) to the possibility of repeated origin of life, and thus also the independent origin of major groups perhaps by spontaneous generation. In the taxonomy of the middle-rank categories he speaks of bi- or polyrheitry. I do not consider it necessary to introduce this new concept. Even if we assume that the phyla arose independently, in the sense of the preceding discussion they would have to be considered monophyletic since there would have been a stem species at the beginning of the evolution of each, regardless of how the stem species arose. They would then simply have to be placed in the system side by side without any interconnection, since the several stem species would not have gone back to an older common stem species.

Animal groups that reproduce exclusively by parthenogenesis present a special case. Such a group is the "order" Bdelloida of the Rotatoria, in which males are unknown. In such groups we cannot speak of "species," and consequently our definitions of the concepts "monophyly" and "phylogenesis" do not apply to them. If we speak of monophyletic groups among them we would have to mean groups descended from one and the same individual, which would be comparable to the clones of the protozoologist. It would be interesting to see how far the methods used to determine monophyletic groups in groups with normal speciation could lead to the discovery of such clones.

Dichotomy and Radiation

The concepts of dichotomy and radiation are also related to the concepts of monophyly and polyphyly and to the extent to which it is possible, with available methods, to recognize monophyletic groups. Strange to say, the view is often advanced that phylogenetic systematics presupposes a dichotomous structure of the phylogenetic tree. Because dichotomy is not the rule, it is said that a system that gives the impression of a continuously dichotomous differentiation cannot be regarded as a true presentation of the actual kinship relations.

If phylogenetic systematics starts out from a dichotomous differentiation of the phylogenetic tree, this is primarily no more than a methodological principle. With the aid of Fig. 62a-d, let us look at its *modus operandi*. Disregarding special cases that are of no significance here, the basic assumption is that every living individual belongs to a particular species, and that this was true during any earlier time of the earth's history. Again disregarding a few special cases, it is further assumed that new species arise only by cleavage of existing species. Consequently existing species must be either the direct descendants of earlier species, or have originated by the splitting of one of these earlier species. The problem of systematics is to determine the sequence in which existing species arose by such processes of species cleavage. It can start from any two species (for example, Fig. 62a: A = *Lumbricus terrestris*, B = *Homo sapiens*). No one can doubt that at some time in the history of the earth there was a species that would have to

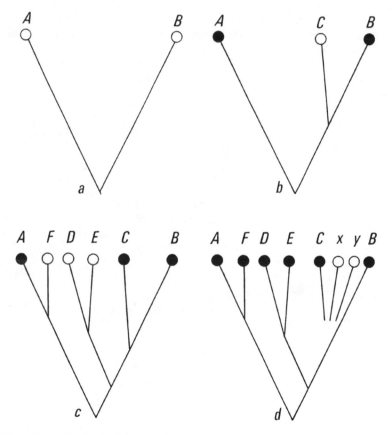

Figure 62. Dichotomy and radiation. For explanation see text.

be regarded as ancestral to both these species. If we now add a third species
(C = *Pongo pygmaeus*), in most cases it will be possible to determine with certainty whether the common stem species of all three species lived before or after
the common stem species of A and B. The result (Fig. 62b) is a dichotomous
system of which no one would argue that it does not represent correctly the relationships of the three species. This procedure can be continued by adding more
species, ending up with a strictly dichotomous system of relatively numerous
species of which again no one would argue that it expresses completely correctly
the relationships of these species (Fig. 62c: D = *Ursus arctos*, E = *Felis leo*,
F = *Musca domestica*). With the addition of more and more species, however,
a point will be reached where the exact relationships of some species cannot be
determined with certainty (Fig. 62d: X = *Pan troglodytes*, Y = *Gorilla gorilla*).
Consequently it will not be possible to erect a dichotomous system for this group
of species (Fig. 62d, B, C, X, Y). But does this mean that the phylogenetic tree
is not dichotomous in this section? It simply means that our methods or the presently available data do not permit a decision as to whether here too a strictly
dichotomous cleavage of several successive species took place or whether there
was a simultaneous splitting of one stem species into several daughter species
(radiation).

Another imaginary example will show that there actually must be cases where
available methods do not permit a decision between these two possibilities (Fig.
63). Species A is distinguished by the characters a, b, c. At time t_1 a marginal
population splits off from A and in the course of time evolves into an independent
species B. During this process the character b develops into b'. Thus during the
time between t_1 and t_2 there are two species, one of which (A_1, with the unchanged characters a, b, c) is indistinguishable from A. At time t_2 another marginal population splits off from A_1 and evolves into the species C, the character c
being transformed into c'. At time t_3 (the present) there are thus three species:
C (with the characters a,b,c'), A_2 (with the characters a,b,c), and B (with the
characters a,b', c). According to our definition of the concept of phylogenetic relationship (p. 91), C is more closely related to A_2 than is B. The species C and
A_2 together form a monophyletic group. Between this and C there is a sister-
group relationship. This could not be determined by our methods, however, since
the relationship would be determinable only with the aid of synapomorphous
characters, which do not exist in our species groups A_2, B, and C. This example
supplies a partial explanation for the fact that it is impossible, for example, to determine the exact kinship relations among the species in many large insect genera.
But above all it shows that the impossibility of determining with certainty the
sequence of dichotomous cleavages in a group never means that all the species
arose simultaneously (by radiation) from one stem species. A priori it is very
improbable that a stem species actually disintegrates into several daughter species at once, but here phylogenetic systematics is up against the limits of the
solubility of its problems. These limits are set by a certain vagueness and inde-

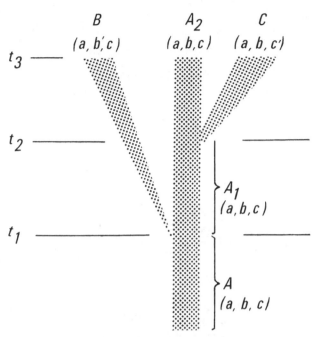

Figure 63. Limits of recognizability of phylogenetic kinship relations. For explanation see text.

terminacy of the concept "simultaneous." On the one hand it is impossible to say exactly when a cleavage process begins and when it ends, and on the other hand indeterminability depends on how long it is before the originally unaltered "daughter species" (A_1 in Fig. 63) becomes altered. If the daughter species A_1 became altered prior to the beginning of the second cleavage process (i.e., before time t_2), for example, by transformation of character a into a', then character a' would have had to be taken over by or further evolved in all later species. In this case exact kinship relations would have been demonstrable.

It is very difficult to get definite information on how long a species can remain unchanged morphologically after splitting off from a marginal population that subsequently evolved into a distinct species of its own. There is probably not a single case in which a species from the Baltic amber is definitely known to agree completely with recent species. Even in the case of *Tetracha carolina* L. (Coleoptera: Cicindelidae), a recent species that is said to occur in the Baltic amber, the specimens from the amber differ slightly from modern specimens, according to Horn. At any rate the differences are so minor that it is impossible to decide whether the Eocene population could have been the stem species of several recent species of *Tetracha*. The picture is similar for other species of insects that occur in the amber. Consequently under certain circumstances we must reckon with very long periods of time during which the sequence of species cleavages is indeterminable. Therefore if the distinction between "radiation" and "dichotomy" is to make any sense at all it can only be that we speak of radiation when the sequence of cleavages within a particular time span cannot be determined.

This raises a question. In younger monophyletic groups it is evidently impossible to distinguish with certainty between radiation and dichotomy, or to recognize a probable consistent dichotomy as such. Should we not then question the clear dichotomy that seems to be recognizable in so many older groups (Acrania-Craniota, Cyclostomata-Gnathostomata, Chelicerata-Antennata, Crustacea-Tracheata, Myriapoda-Insecta, etc.)? May not the picture of a radiation, i.e., an approximately simultaneous origin of several groups, be the correct one for these older groups too, and the picture of a clear dichotomy actually be only an artefact of our methods of recognizing kinship relations? In my opinion this need not be feared. Older groups have gone through many a "time of filtration," with important changes in ecological relationships, and in connection with this many species and species groups that arose approximately simultaneously have perished. Figs. 64 and 65 show how in older groups a clear and appropriate picture of dichotomous differentiation (as inferred from the surviving recent species) may result from the random extinction of species or their descendants that actually arose approximately simultaneously.

So far we have investigated the applicability of the concept "radiation" to the basic process of phylogeny, which is species cleavage. But radiation is often taken to mean not so much the simultaneous or approximately simultaneous origin of species from a stem species, but rather the origin of several stem groups from one "stem group" (Fig. 65). Examples are the origin of mammals, birds, and the modern "reptiles" from the late Paleozoic–early Mesozoic "reptiles," or the origin of numerous groups of Cephalopods from the early Paleozoic Ellesmeroceroida. This is often called a "polyphyletic" origin of the younger groups from the older "stem group," and is a very characteristic confusion of the concepts of typological and phylogenetic systematics. Taken by themselves, the "reptiles" of a particular time —such as the late Paleozoic and early Mesozoic—are certainly a monophyletic group. But if one opposes to them "the groups that arose from them," they become a grouping of the typological system. In such a situation they are no longer a monophyletic group, since the definition of a monophyletic group no longer applies to them: not all species of the late Paleozoic–early Mesozoic reptiles are more closely related to all other species of these fossil reptiles than to certain species outside this group—some are more closely related to the mammals, others to birds, and still others to certain modern "reptiles." This can be called radiation only in the sense of "derivation" of certain groups from a particular type of structural plan, which is something entirely different from radiation in the sense of a phylogenetic process. Nor is it permissable to use the concept "polyphyly" for this, since polyphyly likewise relates to the interpretation of phylogenetic processes and not to the derivability of one or more types of structural plan from one another. Failure to observe this results in the logical error of metabasis.

It seems to me that discussion of the monophyly or polyphyly of the Hystrico-morpha, which play such an important role in analyzing the faunal history of South America (see, e.g., Simpson 1953), suffers extensively from such concep-tual confusion. It is also clearly evident in Simpson's (1959) discussion of the polyphyly of the mammals. To answer the question of whether the mammals are a monophyletic group or not we would have to proceed as described above (p. 88). At least for each of the subgroups (if not for each species) of recent mam-mals recognized as monophyletic, we would have to ask whether they are actually more closely related to other subgroups of the Mammalia than to groups not in-cluded in the Mammalia. The fossils would also have to be considered according to the principles we have discussed. Simpson, however, lists four osteological characters that until recently were considered characteristic of mammals in con-trast to reptiles. He then notes that at least some of these characters arose inde-pendently several times in the history of the Mesozoic "Therapsida," and that consequently several groups crossed the "reptile-mammal line" independently. "The conclusion that the mammals, by structural definition, are polyphyletic is

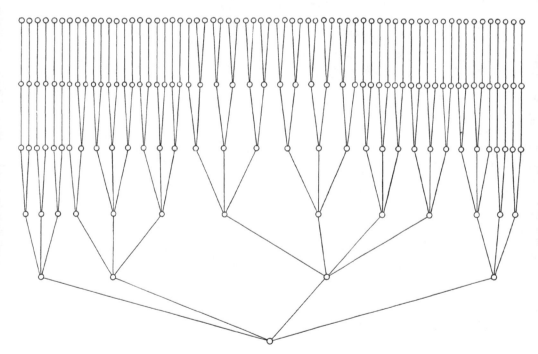

Figure 64. Increase in the number of species of a taxonomic group under the assumption that with in-creasing filling of the space the tempo of species cleavage is reduced more or less uniformly in the younger subsidiary groups. Dichotomic differentiation of the lower categories, polytomic of the higher ones.

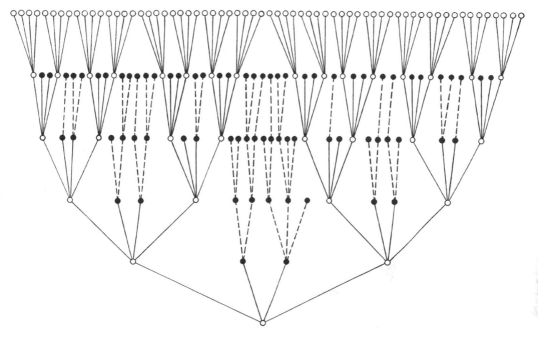

Figure 65. Increase in the number of species of a taxonomic group under the assumption that the subsidiary groups with sharply deviating characters remain the bearers of undiminished rates of species
cleavage at a disadvantage to other (intermediate) suppressed partial groups. Dichotomic differentiation
of the higher categories, polytomic differentiation of the lower ones.

strongly supported by the knowledge of the Therapsida." "The class Mammalia as
currently recognized is thus a grade and not a clade in the terminology of Huxley (1958)." But the facts given by Simpson only show that at least one character (the differentiation of the squamoso-dental articulation), which occurs in
all recent mammals as an undoubtedly apomorphous character, arose independently several times in the Mesozoic, and that this may also be true of still other
osteological characters. But this does not prove that this character or characters
arose independently several times ("polyphyletically" by parallel evolution)
among recent mammals or their immediate ancestors. Simpson has singled out
one of the complex of apomorphous characters distinguishing a group of species
(the modern Mammalia), and then declared its presence among fossils to be
the decisive ("essential") mammalian character, and that any fossil showing this
character—regardless of its ancestry—is to be considered a mammal. The resulting grouping is typological. The conclusion that the group is "polyphyletic"
because this character arose independently in several species cannot be considered reliable. Everything we said above about the paleontological method, the

classification of fossils in the phylogenetic system, etc., should also be considered here. Simpson's study does not prove the polyphyletic origin of the mammals, but rather confirms the fact that a single apomorphous character is very dubious proof of the monophyly of a group, because "parallel evolution" apparently is a very common phenomenon. But this was already evident from what we said above about evaluating the "correctness" of the results of phylogenetic investigation.

Explosive Radiation, Typogenesis, and Related Concepts

Many paleontologists, in particular, are convinced that phylogenesis is typically phaselike. According to Beurlen (1930) the evolution of larger groups—like the life cycle of an individual—is a developmental cycle beginning with their first appearance, followed by a period of differentiation, and terminating in gradual extinction. Two stages are said to be distinguishable in each evolutionary cycle:

1. A stage in which the basic type undergoes explosive radiation into a series of possible structural plans and adaptive types. In each case the structural plans are within the framework of the underlying basic type.
2. A stage in which the new structural plans undergo consistent differentiation. No basically new formations arise that go beyond the framework of these structural plans.

Schindewolf, who embraced the phase theory in the form of his typostrophic theory, distinguished three phases:

1. At the beginning, a saltatory differentiation of the character combination (structural plan) that characterizes the particular stem. Duration very short. Evolutionary tempo rapid. Great structural lability.
2. A phase of long-lasting gradually progressive evolution. Continuous development. Form stable.
3. Overspecialization, labile degeneration and dissolution of form. Extinction.

Evolution is undirected in the first phase; in the second and third phases it is strictly directed, usually in several parallel series (Schindewolf 1942).

Jaekel, Pavlov, Spath, Wedekind, Woltereck, and others have been identified with the theory of phaselike phylogeny. As von Bertalanffy (1942) emphasized, Haeckel already expressed similar ideas in his theory of the epacme, acme, and paracme of evolution.

Romer (1960) has investigated the question of how far explosive evolution is the rule in phylogeny. He uses the term for "an evolutionary process . . . in which a group of a modest sort rapidly attains a prominent position in the fauna without notable modification or diversification of structural pattern." To him it seems "certain that explosive evolution is not an exceptional process but has been for most groups the normal one." Undoubtedly explosive evolution is closely related to what we called "radiation" in the preceding section, at least if by radiation we do not mean an absolute contrast to dichotomous cleavage, but only an accumulated cleavage in relatively short periods of time. Consequently everything we said about the importance of keeping the concepts of typological and phylogenetic systematics strictly apart also applies extensively to the evaluation of explosive evolution and the alleged phaselike course of the evolution of animal groups.

All the examples given by Romer seem to represent monophyletic groups, so we are not looking at phantoms produced by mixing typological and phylogenetic concepts. But Romer also says that "since explosion appears to be generally followed by extinction, animal groups which follow this pattern have played little part in the long-term picture of vertebrate evolution." This is important. Many animal groups have existed since the Precambrian, and investigation would show that a much larger number arose in the older Paleozoic than is apparent from the customary family trees of the animal kingdom. Thus it is in no sense true that all animal groups evolve explosively, and then become extinct. But what groups are there whose evolution does follow this pattern—or perhaps one of the "three main types" in the "major course of the phylogeny of animals" that Müller (1955) believed are distinguishable—and which do not? This question cannot be answered without taking into account differences in form and in their phylogenetic development.

Zimmermann used the term "character phylogeny" for the study of phylogenetic changes in form. It is better to speak of the evolution of form, since the term "phylogenesis" can refer only to "relatives" (see above). But it should not be forgotten that not only the characters or individual peculiarities of organisms must be considered from this standpoint, but also the total form (holomorphy). Rules can be found not only for the further evoluton of individual characters, but also for the total structural plans of systematic units. The quantity of facts that have been amassed on the phylogenetic development of characters and whole character complexes is extraordinarily great, but the area of applicability of these rules is also extraordinarily varied. Besides those that apply to whole phyla or even to the whole animal kingdom, there are others that relate to characters (peculiarities) that occur only in units of very low rank order. The area of applicability of the laws of character evolution does not always coincide with the boundaries of phylogenetic units. Often ecological groups of varied phylogenetic relationships are subject to the same laws. These are usually called "biological rules," and they indicate a close causal relationship between the development of form and the environment. Knowledge of these laws is important for special systematics, as shown in the first part of this book. The problem of adaptation, which deals with the relationship between development of form and the environment, is one of the basic problems of phylogenetics, and special systematics must contribute to its solution. It must be expressly noted, however, that besides these "biological rules" there are laws of phylogenetics that have nothing to do with the relationship between the object and the environment. The phenomena of homoiologies, parallelisms, and trends are among these. These phenomena are limited to relatively low taxonomic groups, but it has also been claimed that there are laws that have universal applicability. Beurlen (1937) mentioned a law of general increase in size during phylogeny. According to Rosenfeld and Goldmann (1939) the activity of the oxidation ferment increases as we go up in the animal series. Backmann (1938) believes it is possible to demonstrate a "multiplication or polymerization of growth cycles" in the course of phylogeny: "without doubt, phylogenetic development is to be understood in many respects as growth. Polymerization of growth cycles can be interpreted as a condition of phylogenetic evolution and the accompanying phylogenetic differentiation." According to Blagoveschenski, cyclic compounds accumulate in the course of phylogeny. The extinction of organisms in the sense of the phase theory—which Blagoveschensky accepts—depends on an overproduction of such ring compounds.

Blagoveschensky (1929) has developed further ideas on phylogenetic hysteresis. He starts out from the peculiar distribution of alkaloids in the plant kingdom; these are found on the one hand in the terminal families of phylogenetic series, and on the other hand in forms retaining archaic and primitive characters that betray early completion of a phylogenetic series. This is likewise true for genera within families in which not all genera have alkaloids. Alkaloids are complex aromatic compounds of very high stability, and consequently their accumulation is a sign of chemical senility, of "phylogenetic hysteresis." The occurrence of large quantities of such stable compounds reduces the reactivity of protoplasm, and limits the possible paths of development—just as, according to Pictet, the death of protoplasm is connected with the stabilization of the protein molecule with attendant formation of ring compounds. Therefore, characterized by the quantity of cyclical compounds, there are "chemically old" and "chemically young" groups. This is probably why the variability of organisms was greater in earlier geological epochs than it is today, because the formerly relatively unstable systems changed, through various impulses from the environment, into more stable and more probable systems. In connection with the accumulation of stable complexes, lability had to decrease, down to complete rigidity and extinction of the species (Rosa's rule of "progressive reduction of variability"). According to Blagoveschensky the evolutionary process tends toward decrease of free energy, i.e., toward the most probable conditions of equilibrium; the most probable systems are the richest in cyclical compounds. In every phylogenetic series we can distinguish stages of youth, maturity, and senility, during which plasticity decreases (von Bertalanffy, II).

From von Bertalanffy's description it appears that the conformities to law assumed by Blagoveschensky, although they relate primarily to a single character in the morphology of the organism, lead over to general laws of development by which development of the whole morphology is said to be controlled. One of these, the "law of progressive reduction of variability," is usually attributed to Rosa, although the same principle was clearly expressed by Fechner as far back as 1873.

The validity of this law is not generally acknowledged. Mez (1926), Plate

(1928), and Naef (1931), among others, have criticized it. The vagueness of the concept of variation used in the law is certainly partly responsible for its varied interpretations. The basic fact of biology, that no living organism is exactly the same as any other, is generally called variability. Consequently in this form of the concept the Fechner-Rosa law would mean that in phylogenetically derived species—i.e., those that are holomorphologically particularly far removed from the initial structural plan—the individual differences would have to be smaller than in "more primitive" species. There is no evidence of this, and personal experience also fails to support it. The "form dissolution" at the end of an evolutionary cycle postulated by adherents of the phase theory (e.g., Schindewolf) likewise speaks against the validity of the law in this sense.

But we can also proceed from our idea of phylogenesis, which says that all evolution of taxonomic groups can be understood only as progressive differentiation. However long phylogenesis may continue into the future, the descendants of existing species of Diptera will always be holomorphologically only extensions of the structural plan of the Diptera, the descendants of the mammals only extensions of the structural plan of the Mammalia. In this sense the evolutionary possibilities (further differentiation, "variation") are undoubtedly more limited than they were for the common ancestors of the Diptera and Mammalia, whose descendants had both the dipteran and mammalian, as well as many other, paths open to them.

"The breadth of evolution of successive groups shows a distinct narrowing, since the basic divergences of organization become progressively smaller. The type of the mammals is more uniform and closed than that of the reptiles, which in turn is unquestionably uniform compared with the type of the Amphibia-Stegocephalia. . . . The same phenomenon is repeated in every systematic unit of higher or lower order" (Beurlen 1937).

This, no doubt, is how the law should be understood, and in this form there can be little objection. But we may ask whether any knowledge has been gained that merits calling it a law. Also the impossibility of framing exactly the concept of similarity and difference in form is evident here. When the last common stem species of the Mammalia and Diptera split up into successor species, the latter would scarcely have differed from one another more than do living species that are related to one another in the first degree. But who can say how much the descendants of the living mammals a hundred million years from now will differ from one another within their still common structural plan, and what yardstick would we use to measure their differences in comparison with the differences among living species?

One thing emerges from all this: the Fechner-Rosa law has inner relations to two other laws of evolution, Dollo's law of irreversibility and Cope's law of the unspecialized. Our assumption that the descendants of the living species of Diptera could evolve further only within the framework of the dipteran structural plan (and that of its families, etc.) is based on the assumption that the law of irreversibility is valid. The area of applicability of this law was discussed in the first part of this book. It undoubtedly has its limitations. With the restrictions that now must be made, it is merely a probability statement concerning the course of evolution of characters: the more complicated an organ the less probable that it can arise again in the same way after it is once lost. This is how Müller (1939) reduced Dollo's law to a probability statement from the genetic viewpoint:

"Thus a very complicated path leads from the original stage to the existing type, so it is scarcely possible that the original condition can be restored by a back mutation of the main gene. For smaller and very young phylogenetic changes, however, reverse developmental stages may occur and even be observable" (quoted from Wettstein 1941).

Like Dollo's law, Cope's "law of the unspecialized" (or "law of primitive stem forms" as Plate [1928] prefers to call it) is related to the Fechner-Rosa law of progressive reduction of variability. If the possibility of producing new structural plans decreases progressively—with progressive removal from the origin of the pattern of the stem form—then it is understandable that new types that persisted in the further course of evolution often arose from stem forms morphologically similar to the original form. Cope's law leads to the problem of the relation between species cleavage and change in form, which is discussed below.

Abel (1929) gathered Rosa's, Dollo's, and Cope's laws, together with the "law of orthogenesis," under the general term "law of biological momentum." It cannot be denied that there are internal relations among them, but it is doubtful whether there is the slightest gain in Abel's proposal, especially since there is nothing to justify the analogy to physics implied in the name.

All the laws so far mentioned belong to the circle of problems that we have called "character phylogenetics," i.e., they are statements regarding conformities to law in the evolution of individual configurations either as wholes or in their parts. We may ask whether there are not also laws that determine the course of evolution as a whole, whether for example phylogenesis moves toward a particular goal or even several goals. Whether we can speak of a law if this question can be answered in the affirmative must remain uncertain, since the action of a law is not restricted to a particular case whereas phylogeny as a whole is a nonrepetitive process.

The most general evolutionary trend, the tendency toward gradual perfection, or "higher development," has been called a law. As always in such cases, these two concepts are in no sense unequivocal, and have not always been used in the same sense. It seems to me that the only possibility of speaking of a "higher development" in phylogenetics is to give this concept, with von Bertalanffy, the sense of an "elevation of the stage of organization," an increase in internal differentiation, the "achievement of a higher level of integration" (Friederichs 1937). The concept of "perfecting" would have to be separated from this; it can be related only to the possibility of the organism dealing successfully with its environment, as formulated by Plate (1928) and not always consistently by Franz (1935).

We then face the problem of investigating the relation between higher development and perfecting. Perfecting can be explained by the four known evolutionary factors. Only if it can be shown that higher development is not always connected with perfecting—much more difficult to prove than many theoreticians of phylogenetics seem to think—could we see a general law of phylogenetic evolution in the tendency toward elevation of the stage of organization.

So far we have dealt exclusively with the form of organisms, which in general language usage we call individuals. We must now recall that species and higher taxa are in a certain sense also individuals in the phylogenetic system. If higher development by elevation of the stage of organization, increase in internal differentiation, can at least be considered as the goal of phylogeny, we must ask whether this development relates only to individuals in the usual sense, or whether the individuals of higher order do not have a form (species, genus, family, etc. form) capable of higher development. Woltereck (1940), for example, regarded the origin of species and higher taxa as an expression of a general tendency toward diversification (anamorphosis) in the course of phylogenesis, but this is not a positive answer to our question. Mere diversification in species and higher taxa is not the same as the origin of new specific, generic, and family structural patterns—which must not be confused with the "type" (the average, so to speak) of these groups. So far as I know, Fechner (1873) was the first to consider the possibility of regarding the higher taxa as the possessors of individual group forms.

Paleontologists in particular have urged that the superindividual "form" of higher phylogenetic units be worked out. Marinelli (1939), for example, proposed making "the total form of an animal stem the object of morphological consideration." He wrote:

The main stem of life ascends in another sense, which we best understand by comparison with the morphology of the individual organism. In the latter, development is by progressive internal complication, the parts always differentiating in different directions through division of labor, and thereby becoming the more intimately tied together into a higher unit. Similarly in phylogeny there is first a breakdown of diversity —progressively more sharply separated special types develop which, taken together, build up (precisely by their division of labor) an ever more narrowly closed total fauna of a living space, a time period. These are considerations that no longer work with the individual developmental series and the arrangement of forms within them. Rather the object is the stems themselves, and from their development to discover laws of living morphosis. This would be morphology of the stems.

Von Huene (1940) expressed himself similarly:

Thus it seems to me that the total evolution of life through the ages is pervaded by a superordinate principle that orders the totality in a uniform and direction-giving sense (cf. Spemann's organizers, on a small scale), and which acts cooperatively in all possible orders of magnitude. In this way there arises a four-dimensional total form (time being the fourth dimension) of the totality of organisms, much as the individual gradually develops all organs by growth from a single cell. The organs do not develop in a directionless way; they develop within the framework of the whole, and some disappear before the body as a whole reaches maturity. As the particular is dominated by the uniformity of the whole in the development of the individual (which in turn is only a subordinate part of another whole), so does the total evolution of organisms also show a higher organization, an organic course of life.

Attempts to grasp the total form of higher taxa—not to be confused with the "type" of idealistic morphology—have scarcely gone beyond the programmistic character expressed in the statements quoted above. One of the first positive observations is that of Ziegenspeck: "Characters that in phylogenetically . . . lower groups still occur within close circles of kinship, in the derived forms become so specialized that they are inviolable family characters" (quoted from Mez 1926). If this proves to be generally correct we would have in the evolution of higher groups the same principle seen in the development of the individual: peculiarities originally belonging to a single cell are distributed among the descendant cells which thus become distinguishable as a group from other cell neighborhoods that have received a different set of characters. The essential thing in this development is not at the morphological level; it is that the functions dependent on these peculiarities supplement each other in such a way that the organism becomes a truly functional unit within its environment. Only if the same could be proved for the higher taxa, with a differentiation and mutual augmentation of functions connected with the "related differentiation of structures" to make the total group an action unit, would a comparison between the higher taxa and individual organisms be more than a loose analogy.

It must be admitted that a mutually related differentiation, not only of structure but also of functions in the environment, is probable in all cases where the species-number curve follows the logistic equation of population growth. With simultaneous origin of differences in form among species, a species-number curve of this form would scarcely be understandable if the species differed from one another independently. All these questions lead far beyond the circle of problems we are discussing here. They are mentioned only because they deal with fundamental questions of evolutionary research that cannot be solved without the cooperation of the special phylogenetic systematics we are primarily concerned with here.

The laws of the development of form that were discussed briefly above are not by themselves important to special phylogenetic systematics. They are important because, following the principle of reciprocal illumination, they can help disclose the phylogenetic kinship of related groups, which is the primary task of special systematics. As already emphasized, this is possible with the aid of analysis of form only if orderly relations exist between directly recognizable alterations of form and those on which differentiation of the species—and thus of the "groups" in general—is based. The nature of these relationships will have to be investigated a little further. The basic fact is that there is no definite correlation between magnitude of morphological difference and degree of phylogenetic kinship, although rules with a certain range of applicability can be found. Equally important is the fact that there is no such thing as specific, generic, family, or ordinal characters.

This opinion is not shared by many modern phylogenists, who believe there is an essential difference between higher and lower taxonomic groups and the attributes that characterize them. Some, for example, distinguish between constitutive and additive characters (Woltereck 1940), constitutive and "nonconstitutive or functional" characters (Diels 1921), adaptive and organization characters (Schindewolf 1943, Stubbe 1942, and Wettstein 1942), or even between "type-caused fabric characters" and "adaptively caused special characters" (Schindewolf 1943).

The type doctrine of idealistic morphology lives on this distinction between two supposedly fundamentally different processes of evolution: aromorphosis and adaptiogenesis (Sewertzoff), eogenesis and evolution in the narrow sense (Clark), micro- and macroevolution, typogenesis and adaptiogenesis (in the sense of Jaeckel, Philipschenko, Vialleton, Osborn, Schindewolf, Beurlen, Schuh, and others, according to Heberer 1943), cladogenesis and anagensis (Rensch, see Simpson 1949).

Naturally this idea is extremely important to our concept of the current state and the future of the evolutionary process. "It is not true that phylogenesis can continue indefinitely. Rather it seems to have reached its material end, since today there are no longer any unspecialized and phyletically fully active groups, as paleontological history shows" (von Huene 1940). "Cladic evolution . . . ended long ago." No "new structural type" has appeared for more than 600 million years (Cuénot 1940). Woltereck considers it "absolutely inconceivable how new structural plans could branch off from the presently dominant type circle of the vascular plants, the mammals, or the insects." Like Cuénot, he believes that "the production of new somatic types apparently has ended in both the plant and animal kingdoms." "There is no evidence whatever that a future origin of correspondingly different types is possible, and much speaks against it."

On the one hand this view leads to pessimistic ideas regarding the future of phylogenesis: "The axis is dead; evolution continues, of course, but within the clades, of which only the summit is alive" (Cuénot 1940). On the other hand, it necessarily leads to skepticism regarding all attempts to explain—by genetics, for example—evolution or the essential processes of phylogenesis from phenomena observable today. According to Schindewolf (1942), "the first representative of a new stem, at the ordinal level for example, does purely formally represent a species. . . . But this 'species' monotypically embodies at the same time the genus, family, and order, and is potentially completely different from the species that later emerge from this stem, which are the terminal members of the differentiation." Everything genetics has learned about the mechanism of speciation is said to relate to the latter kind of species, whereas the origin of the stem species of higher taxa through the evolutionary mechanisms known to geneticists is unintelligible. Among the zoologists, particularly geneticists, who at least do not deny this idea, Schindewolf (1942) names Remane, von Buddenbrock, Ludwig, and H. Ulrich; Goldschmidt can be added.

Representatives of an even more extreme type doctrine go so far as to deny phylogenesis altogether, or at least interpret it as of very subordinate importance within the narrowest framework. "The origin of the plant and animal kingdoms is polyphyletic. It begins with genera and is bound to genera. There is no phylogenesis in the mechanistic-monistic sense." "As monism must lead, by way of the monophyletic theory, to spontaneous generation of the primordial cell of the entire plant and animal kingdom, so must holism necessarily lead, by way of the polyphyletic idea, to genogenesis or to primordial germs of the genera" (Kempermann 1936). Kleinschmidt (1926) earlier presented similar and, if possible, even more extreme views.

In trying to determine our own position on this important question we must note first of all that we do not believe showing how species arose, or at least could have arisen, is a sufficient explanation of the evolutionary process. It no more explains the origin of higher taxa than does the origin of mutations in certain individuals explain the process of speciation. The aim of this whole book is to develop this fact as clearly as possible. But this is not the decisive factor in the position of those upholding the type doctrine. They contend that in speciation— or even in the initial process of altering the form of an individual—there are already decisive differences depending on whether the species is destined to be the stem species of a new "type" or merely one of the differentiation products of an old stem.

If it were possible to determine by objective criteria whether the daughter species arose by cladogenesis or by anagenesis, or whether at least one of them is destined to be the stem mother of a "new type," then our diagram of hologenetic relationships (Fig. 6) would in fact be incomplete or even misleading. Instead of "phylogenetic relationships" we would then have to use two different kinds of interspecific genealogical relationships, and it would make little difference whether we distinguish them as cladogenetic or anagenetic or whether we used some other terms. Obviously everything else would then be useless, including what we have said about the structure of the phylogenetic system, and particularly the principle that sister groups must have the same rank.

It is very important to recognize that the typologists do not tell us what they actually mean by "constitutive characters" or a new type characterized by them, and when we are justified in speaking of a new type in a systematic group. Schindewolf (1943) defines the characters of organization as "peculiarities of form that characterize units of higher rank from genus upward," and therefore in this sense he agrees with Kempermann. Other authors restrict the type concept to the highest phylogenetic units: "eogenesis is simultaneous formation of the basic forms in the animal phyla, evolution the continual change that from age takes place within the phyla" (Clark). Most authors give only very indefinite statements. In the presentation of Dacqué (1935), for example, we look in vain for a concrete clue as to when we are justified in speaking of a new type. Radicalization of contrasts is probably a useful tool, and it is easy to work with extreme examples. But anyone who has worked with the systematics of a large modern group will have to admit that it is impossible to decide with certainty between type-related groups and those that do not qualify as such. It is a general phenomenon that in groups that have been worked over intensively by specialists the subgroups are gradually elevated in rank: species groups become genera, and finally separate families. This is no doubt because the type-related characters of even the smaller units become increasingly evident to one who works with them intensively. The distinction between independent type and mere pronounced form within the framework of a "type" is to a considerable extent more a matter of point of view, of the perspective of the investigator, than of objective fact. That this is so, and that consequently the rank order of systematic groups cannot be defined on the basis of morphological criteria, is the reason why we looked for other methods of achieving this goal. And how is it that larval and imaginal forms differ in the expression of structures pertaining to the type? Schindewolf would certainly recognize the Diptera and Coleoptera as separate types with their own constitutive characters. But this would not happen if only the larvae were considered, since with a given larval form it can be very difficult to decide whether we are dealing with a dipteran or the larva of some other insect. On the other hand some subgroups of the Crustacea are shown to be crustaceans more definitely by having nauplius larvae than by the organizational characters of the adult stage.

But even if we admit that there are groups in which the structural plan differs from the structural plans of related groups, this does not mean that a special law must have governed the origin of such groups. If we equate the somewhat ambiguous expression "macroevolution" with "origin of higher categories," then this would agree with Nachtsheim's statement that "the laws governing macroevolution are the same as those governing microevolution." Also according to Heberer (1958) the distinction between macro- and microevolution corresponds to "no reality." "Genetics is not in a position to support in any way typostrophism, the type-discontinuity doctrine, the systemic mutations of Goldschmidt, or the saltations of many paleontologists." Even those who insist that these views are not generally accepted must admit that the phylogenetic system is of decisive importance in clearing up the question. Its job is to refrain from using typological criteria of any kind in determining the rank of higher taxa, limiting itself to a clear expression of sister-group relations. Only after this has been done can we investigate whether or not special conditions are associated with the origin of groups that—in comparison with their sister groups—have the character of a "new type," a "new plan of organization," a "new anagenetic stage," or the like.

Obviously if erroneous conclusions are to be avoided in discussing the mode of origin of "higher categories" it must be sharply distinguished whether the higher categories of the phylogenetic or some typological system are meant. In the latter case we must also know the criteria used in evaluating the rank order of the categories in that particular typological system. The principles and generalizations of Simpson (1959) show how important this is: "Higher categories generally arise by acquisition of a basic general adaptive complex, which may be retained essentially unchanged in all subsequent members . . . or may be profoundly modified or lost in some. . . . The origin of a higher category may involve at one extreme (e.g., rodents, birds) a single character or adaptive complex or at the other extreme (e.g., primates, mammals) general adaptive change or improvement in numerous ways or in virtually the whole organization."

In my opinion these generalizations say nothing about the mode of origin of the "higher categories" in the sense of phylogenetic systematics. They merely set forth how higher categories with a particularly outstanding type-connected character tend to differ from their sister groups. What about the "basic general adaptive complex" of the Archosauromorpha, which are a higher category in the phylogenetic system (but only in this system), and in fact a higher category than the Aves, or the Theromorpha, which are a higher category than the Mammalia? Do Simpson's generalizations apply to these? One gets the impression that a few particularly high categories are characterized only by a few unimportant features of their basic plan—as for example the Deuterostomia by a few peculiarities in the ontogenetic mode of origin of the intestinal opening and mesoderm. It would seem that such peculiarities did not arise as an "acquisition of a basic general adaptive complex," and that they are constitutive characters of the basic plan of a higher category only because they were retained in all descendants of the stem species that evolved them (perhaps because they had little adaptive value and therefore were forgotten by evolution). On the other hand we could postulate that all higher categories must have been characterized by special adaptive qualities at their time of origin, because otherwise the characters would not have been retained for such a long time. This, however, can scarcely have been the sense of Simpson's statements.

In any case it seems certain, to me at least, that the only way to achieve confirmed concepts on the mode of origin of the higher categories is to compare as many cases as possible, particularly those that arose at different times and under different circumstances. Only the phylogenetic system can do this—without preselecting them, it renders all recognizable segments of the phylogenetic tree (i.e., all higher categories) uniformly specified and comparable. The typological or syncretistic system cannot do this—from certain points of view it makes a choice among the higher categories and places those that are provided with type-related characters in the foreground. Investigation of the systematic "fine structure" may help to understand the development of those higher categories that at first sight seem to be sharply circumscribed, and which consequently may be suspected of owing their existence to a particular mode of origin. It is evident, for example, that an increasing number of insect groups have a form of differentiation like that shown for the Lepidoptera in Fig. 66. The kinship (sister-group) relations shown in this diagram are probably accurate in all essential respects, even if it turns out that one group or another—e.g., the Aplostomatoptera—should be given a somewhat different position. The diagram merely shows that sister groups differ greatly

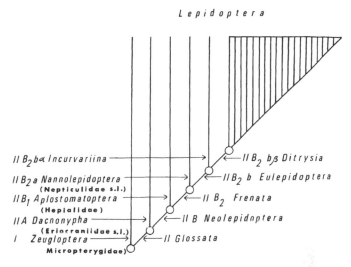

$Lepidoptera$

$II\,B_2\,b\alpha\,Incurvariina$ ——→ ←——$II\,B_2\,b\beta\,Ditrysia$

$II\,B_2\,a\,Nannolepidoptera$ ——→ ←——$II\,B_2\,b\,Eulepidoptera$
(**Nepticulidae s.l.**)
$II\,B_1\,Aplostomatoptera$ ——→ ←——$II\,B_2\,Frenata$
(**Hepialidae**)
$II\,A\,Dacnonypha$ ——→ ←——$II\,B\,Neolepidoptera$
(**Eriocraniidae s.l.**)
$I\quad Zeugloptera$ ——→ ←——$II\,Glossata$
Micropterygidae)

Figure 66. Principal differentiation of the butterflies in the phylogenetic system. Because of the distribution of the species numbers in the subsidiary groups there results the impression of a directed phylogenetic development (see text).

in species number. But it is also true for the differentiation of structure that the complex of characters present in the main mass of the Lepidoptera (Ditrysia), and which so strongly determine the type-related character of the group as a whole, is reached only by stages. Each of the species-poor subgroups entered on the left in the diagram lacks some of the characters of the total complex (see the statements in Hennig 1953, Fig. 9, which are somewhat in need of revision but still essentially correct). This does not completely bridge the morphological gap separating the main mass of the Lepidoptera from the sister group of this "order" (the Trichoptera), to be sure, but the idea that this gap did not arise by a single evolutionary jump is substantially supported by the presence of the relatively plesiomorphous subgroups of Lepidoptera and the graduated kinship relations between them. The limited number of species in these groups also suggests that the still more plesiomorphous sister groups needed to bridge the gap once existed, but have become extinct. The significance of everything that was said above regarding the importance of the distinction between age of origin and age of differentiation is also emphasized.

The Insecta as a whole have a systematic structure similar to that of the Lepidoptera (Fig. 59). The character complex that decisively determines the general idea of the "type" of the insects is reached only with the Pterygota. The differences between the basic plan of the Insecta and that of their sister group (the Myriapoda) are relatively slight. It does not seem that the apomorphous basic-plan characters of the insects had a particularly high adaptive value at first, since the number of species in the groups that do not have much more than the derived basic-plan characters of the insects is extremely small compared with the total number of the Insecta. That these groups survived to the present at all is probably only because they saved themselves by special adaptations (which have little to do with the basic-plan characters of the insects) to particular ecological niches. The basic-plan characters that arose first (heteronomy of segmentation, loss of abdominal limbs, and a few others) became important only after the origin of wings. The essential preliminary conditions for the latter were likewise reached in stages, as is clearly shown by the relatively plesiomorphous, still flightless subgroups of insects that have been preserved. In monographic presentations, especially in textbooks and handbooks, the character complexes that determine the type-related character of the higher categories (such as the insects) are usually overemphasized, whereas the preserved pre-stages, if mentioned at all, are dismissed in a few words as "exceptions to the rule." Only the phylogenetic system provides the proper place for them and permits clear recognition of their importance in phylogeny.

The theory of the "early ontogenetic origin of types" assumes a special place in discussions of the origin of higher categories or new types. Equivalent or nearly equivalent are the concepts of "proterogenesis" (Schindewolf), paedomorphism (Garstang), and—as Wettstein 1942 emphasizes—the designations diametagenesis (Mijsberg), fetalization (Bolk), and neomorphosis (Beurlen). We must mention this theory here because, under certain circumstances, it could be important to the interpretation of the morphological methods of phylogenetic systematics. De Beer (1959) formulates the content of this theory: "Evolution is not confined to modifications of the adults, as Haeckel thought, but can start from modifications in young stages and, when it does, its results are the most striking." He explains that the concept of paedomorphosis is designed "to account for those cases where instead of the young resembling the adult of the ancestors, as Haeckel insisted, the adult of the descendant resembles the young of the ancestors."

From the examples given by de Beer it seems to me that two different phenomena are involved, and that these have a very different importance in phylogenetic systematics.

One phenomenon is illustrated by the gastropods. One of the apomorphous characters of this group is the torsion of the visceral sac. According to Garstang this character first appeared as a larval character of the veliger stage and was not transferred to the adult stages until later. This phenomenon would present nothing basically new to phylogenetic systematics. We showed above (p. 123) that phylogenetic systematics must reckon with the possibility that under certain circumstances monophyletic groups can be recognized as such only in certain stages of metamorphosis. If torsion of the visceral sac were the only apomorphous character of the Gastropoda (which it is not), as long as the character was restricted to the larval stage the group could have been recognized as monophyletic only with the aid of these larval stages. Not until the character was transferred to the adult stages would the monophyly have been recognizable from them too.

The other phenomenon, likewise called "paedomorphosis," for which the "derivation of insects from larval myriapods" may serve as an example, has an entirely different significance for systematics. To understand this example we must remember that the term "larva" is not completely unequivocal. On the one hand, free-living juvenile stages with characters that are not immediate pre-stages—and which are therefore called "larval" or "nonimaginal" characters—are designated "larvae." The above-mentioned veliger larvae of the Gastropoda are larvae of this kind. A character of this kind would be the torsion of the visceral sac. On the other hand, free-living juvenile stages that do not have such larval or nonimaginal characters are also called "larvae." Such larvae differ from the adult (imaginal) stages only in that certain characters have not yet reached the imaginal condition of expression. The larvae of the myriapods belong in this category. The presence of only a few (perhaps three) pairs of legs, in which they differ from the adult stages, is based on the fact that the remaining pairs of legs have not yet developed. Consequently the assumption that the insects arose from myriapod larvae is meaningless. It may mean—if there are no nonimaginal characters for myriapod larvae as a whole, or for the larvae of some subgroups of myriapods—either that the insects arose from the stem form of the recent Myriapoda, or that the larvae of any more or less subordinate subgroup of the Myriapoda became insects. No decision between these different possibilities can be reached with morphological methods. If paedomorphosis of this kind could be made responsible to a noteworthy extent for the origin of the higher categories, then the usability of morphological methods for discovering sister-group relationships would be very much limited. In my opinion there is no adequate basis for assuming that the second kind of paedomorphosis played a role in the origin of the insects or other groups mentioned by de Beer. This question cannot be resolved as yet. Moreover it is an example showing that not only does the phylogenetic system itself have "true insight value," but also that "new results as well as new problems constantly arise" from the theory of phylogenetic systematics (Stammer 1961). The very different significance of the two phenomenal forms of paedogenesis for the usability of morphological methods in the recognition of phylogenetic kinship ties is clearly shown in the accompanying diagram (Fig. 67).

PHYLOGENESIS AND SPACE

The idea of allopatric speciation is almost a dogma today. In the structure of the lower taxa the importance of geographic isolation is expressed in the vicariance of the most closely related groups (subspecies). An important question, still not completely settled, is the extent to which in small spatial relations the purely geographic dimensions can be separated from the ecological dimensions of the environment. On this depends the interpretation of situations (particularly seen among parasitic and phytophagous insects) where stenoekous species live in a

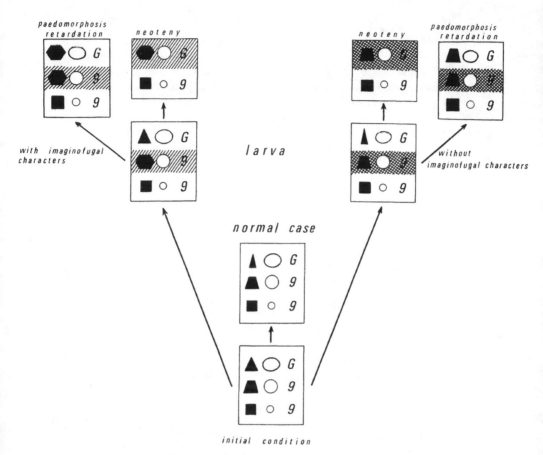

Figure 67. Neoteny and paedomorphosis. The middle and left rows of figures in each rectangle symbolize the ontogenetic development of two different groups of characters. Right: maturation of the gonad (g, immature gonad, G, mature gonad). Acceleration and imaginifugal deviation or retardation of the left group of characters (in comparison with the middle standard group) or acceleration of sexual maturity results in different forms of neoteny and paedomorphosis which are of different significance for the phylogenetic systematics (see text).

limited space in the middle of the range of widely distributed euryoekous species. The importance of purely geographic-spatial isolation in speciation is beyond question. Consequently it would be astonishing if it were not a factor in the evolution of whole groups of species, and were not reflected in the structural picture of higher categories. Actually, vicarious relations between sister groups of certain higher taxa can often be demonstrated (p. 138). This should particularly be the case if the geographic isolation involved in the origin of sister-group relations resulted from the interruption of the connection between extensive areas of the earth's surface, such as entire continents, or where individuals or groups of individuals succeeded in reaching continents difficult to get to and then became the stem species of higher categories. At any rate, the extent to which changes in the overall differentiation of the earth's surface influenced the phylogenetic evolution of organisms is an important question. It has often been tried, without real success, to demonstrate relationships between orogenic processes or similar geological events and phylogenesis, and "clear, simple connections to evolutionary history with so-called diastrophic rhythms are not supported by unambiguous evidence" (Henbest 1952). "In all probability, the fluctuations in evolutionary activity are not directly related to times of widespread orogeny, and there is no indication that peak times of evolutionary activity correspond with times of extensive orogeny" (Newell 1952). Opinions are divided over the importance of the changing connections between continents and continent-like areas. "The main pattern of dispersal of vertebrates is apparently evolution of successive dominant groups in the great, favorable areas of the main part of the Old World tropics and spread into smaller and/or less favorable areas with successive replacements" (Darlington 1958). According to this it is perhaps necessary to think of small areal isolation in the Old World tropics, insofar as isolation participated in the origin of the main groups of the vertebrates.

A very different picture is sketched by Jeannel (1950) for the evolution of insects. According to him, different stocks of insects evolved during the Carboniferous, on the one hand on the then equatorial Laurentian continent, and on the other in the temperate or cold Gondwanaland. At the end of the Permian an equalization took place, favored by the retreat of the seas. During the Mesozoic again a number of different groups arose on the Gondwana, Laurentian, and Angaric continents. This picture is rather sketchy, and does not permit recognizing which sister-group relationships are assumed to have arisen by temporary isolation of these old continental areas. Probably it is not yet possible to test the reliability of such broad hypotheses. This example does show, however, that it would not be correct to interpret the question of the importance of major changes in the earth's surface solely from the relations presented by the vertebrates.

If sister groups of higher order vicariate, i.e., represent each other in different continental areas, it does not always follow that the earlier connection and subsequent isolation of these areas is primarily responsible for the origin of the sister-group relations. The separation of sister groups, one of which is distributed over the northern continents and the other in South America, may perfectly well have originated in North America through small spatial isolation. The distinction between age of origin and age of differentiation of monophyletic groups is always important in clearing up such questions. In general it can probably be said that demonstrating a sister-group relationship between two monophyletic vicariating groups is not by itself adequate proof of hypotheses as to the origin of the vicariance. For example, proof of a simple sister-group relationship between an Australian–New Zealand and a South American group does not permit a statement regarding the mode of origin of this vicariance relationship (e.g., by inter-ruption of an earlier Antarctic land connection) (Hennig 1960). This, like the question of the mode of origin of new "types," requires a more detailed investiga-tion of the systematic structure of such groups.

It is commonly assumed that the geographic distribution of characters provides information on the history of the distribution of an animal group. If, for example, the plesiomorphous species are found in one part of the total range and the apo-morphous species in another, it is assumed that this group spread from the area of the plesiomorphous species into the area of the apomorphous species. This hypothesis is probably justified "in kinship groups with an undisturbed history of dispersal and evolution, that is in species whose systematic structural picture is not torn by great gaps resulting from the extinction of species" (Hennig 1960). Monophyletic groups in which the distribution of characters can be placed in relation to systematic differentiation—i.e., to the hierarchy of systematic relation-ships—always give the impression of a directed progression. If for the series of transformations of a certain character (a, a', a", a''', etc.) it can be shown that the relatively plesiomorphous stage of expression occurs only in the basic plan of one sister group and the next higher apomorphous stage in the basic plan of the other, and if this picture is repeated in the subsequent subordinated sister groups, then the impression of a directed character progression is produced. If a similar distribution of the transformation stages of the same character can be demonstrated in closely related groups, a picture of parallel evolution results. Consequently the possibility of recognizing directed character evolution (ortho-genesis) and parallel evolution in recent animal groups, without knowledge of their ancestry, is bound to a knowledge of the exact kinship and sister-group relations—that is, to the phylogenetic system. If now a directed character dis-tribution of a species can be connected with a likewise directed distribution of the ranges of the bearers of these characters, then there is a basis for the hypothe-sis that the evolution of the character was connected with the geographic distribu-tion of the group. I have called this hypothesis, or the phenomenon it attempts to explain, the "rule of progression" or the "law" of parallelism between morpho-logical and chorological progression (Hennig 1950). Chorological progression could also include distribution in the environment: according to Szidat (1956) "In aquatic turtles an entire evolutionary series of trematodes of the subfamily Pronocephalinae can be erected; the site of occurrence is gradually displaced from the stomach toward the rectum and urinary bladder. This displacement produces a distinct advance in definite directions among the members of this series, expressed in a shifting of the testes into the posterior part of the body and a stronger notching or ramification of the sex glands and gut as a character of successive genera." The rule of progression has special importance in zoogeo-graphic investigations.

In an investigation of the dipteran fauna of New Zealand I developed, on the basis of this rule of progression, a number of ideas as to how arguments supporting an earlier land connection between South America and Australia–New Zealand would have to appear (Hennig 1960). The scheme of one of these arguments is shown in Fig. 68. All have as a prerequisite an exact knowledge of the systematic fine structure of the groups that are distributed in both areas. So far it has been impossible to decide whether, using the Diptera, any of these arguments justified the hypothesis of an earlier Antarctic land connection. This is essentially because the principles of phylogenetic systematics have not been applied accurately enough to any of the numerous pertinent subgroups of this order of insects.

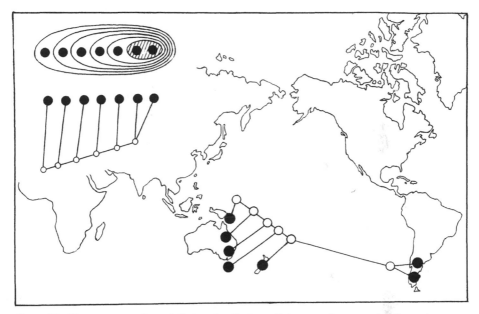

Figure 68. The occurrence of a relatively subordinate partial group of a monophyletic species group that is otherwise distributed in Australia–New Zealand, as argument for the former existence of direct connections between these areas.

IV

Concluding remarks

Much that is controversial and erroneous has been written on the questions of whether a phylogenetic system is possible or necessary, what is meant by phylogenetic systematics, and whether a system in which the phylogenetic viewpoint is combined with others might not be best.

Many of the misunderstandings and differences of opinion could probably be eliminated by not starting from the question of what the present system of organisms—or at least of animals—is and what it should be in the future. A much better point of departure is to recognize that evolution is a fact and that its course and the conformities to law that control it must be investigated. Then we would probably reach extensive agreement that it is the existence of species that gives the evolution of organisms its particular character. Species are important not because they are more or less sharply circumscribed morphological units, but because normal reproduction takes place only within species boundaries. This in turn is essentially based on the genetic peculiarities of the species. There will be no severe differences of opinion on this point. Differences of opinions do flare up over the fact that, in practice, species boundaries are not determined directly by means of genetic criteria, but indirectly by morphological criteria. Consequently even today it is often maintained that the species is a morphological unit, and this gives rise to a misunderstanding or a logical error that drags through many discussions of the merit or lack of merit of phylogenetic systematics in comparison with morphological systematics. These involve the confusion of the practical plane on which we try to grasp particular individual phenomena by means of accessory criteria and approximation methods with the theoretical plane on which these phenomena are ordered into a general relationship.

The assumption that new species arise by cleavage of existing species will scarcely be disputed. There are, however, likely to be differences of opinion over the question of whether a species that changes without splitting should be called a "new species" after a certain point in time. In our context this is an unimportant side issue. Decisive is the fact that processes of species cleavage are the characteristic feature of evolution; they are the only positively demonstrable historical processes that take place in supra-individual organism groups in nature. Through them evolution becomes "phylogenesis." Even those who maintain that there is a special "typogenesis" would not deny that "new types" arise first of all as "new species."

In order to understand phylogenesis it is necessary not only to investigate accurately the processes of speciation taking place before our eyes or what we can interpret as different stages in speciation processes that are actually taking place. We must also try to get as accurate a knowledge as possible of the speciation processes that took place in the past, and which in their totality are to be made responsible for the condition in which organisms—or at least the animal world—is today. This involves above all a knowledge of the sequence in which existing species arose by the cleavage of earlier species. By this sequence we measure the degree of the "phylogenetic kinship" of species. The results are presented by arranging the species in a hierarchic system. Thus "phylogenetic systematics" means no more than the essence of all attempts to determine the phylogenetic relationships of species and to express these in an unmistakable manner. There can be many differences of opinion within the framework of this intent: over the reliability of methods, the applicability of the concept of "phylogenetic relationship" in different animal groups, over the results in detail, and perhaps even over the suitability of the mode of presentation. All these differences of opinion remain within the limits of one and the same theory of phylogenetic systematics, provided there is agreement on the definition of "phylogenetic relationship." If this concept is defined differently, then discussions of individual questions are useless and misleading. If we agree that answering many questions regarding the course of phylogeny requires an accurate knowledge of the phylogenetic kinship of species and an unmistakable presentation of this knowledge, this means that a "phylogenetic system" must be created to the extent that it does not already exist.

We already have a "finished" system of all known animal species, although it is not presented anywhere in its entirety but is scattered through countless individual papers. This is a hierarchic system, but not all of its parts express the phylogenetic kinship of species. For some parts of the system there are no studies that make any attempt to express phylogenetic kinship. For other parts of the system there are several "constructions," some of which actually express the phylogenetic kinship of the species whereas others express morphological similarities by measuring according to some method the "magnitude" of the correspondences or differences. Still others try to express the boundaries between "essentially" different structural plan stages.

The often-discussed question of whether systematics should erect a phylogenetic, a typological, or a combined system can actually only be formulated as follows: Should systematics, if "constructions" are available for parts of the system, decide in favor of a phylogenetic or of some other construction, and should the systematist in the course of reworking parts of the system—i.e., in the monographic presentation of particular animal groups—try to erect a consistent phylogenetic system because it is preferable to others, or are there reasons that make other constructions equally desirable or even preferable? The answer is obvious: anyone who must have a phylogenetic system for particular investigations—there are problems for which this is absolutely necessary—will use only those parts of the system that are constructed according to the principles of phylogenetic systematics. He would have to disregard other parts of the system. If we cannot agree that in the future systematics should be pursued only as phylogenetic systematics, or if we believe that other systems are also necessary because only comparison of these different systems can lead to particular insights, we should nevertheless state clearly in any systematic work on what base it rests. If a work is based on the phylogenetic system, then the foundation and presentation of the system it erects are subject to the requirements set up by phylogenetic systematics and the interpretations stemming from these.

The fact that a consistent phylogenetic system is indispensable for comparative investigations aimed at solving broad problems regarding the course of phylogeny and its laws defines our position on the proposals—often well meant—of a compromise between phylogenetic and typological systematics. Any systematic work, and any work aiming at a consistent phylogenetic system, must necessarily lead toward a compromise. This corresponds to our idea of the "infinite task" of phylogenetic systematics—as of any other science—whose solution can only be approached asymptotically. From this it follows that any compromise with which a work is broken off before reaching its goal should be a forced and provisional compromise. Following what we said above, each author is obliged to make clear where in his work he was forced to make such a compromise and how it is expressed in the system he erects.

Let us take the example of the Psocodea that was discussed above. In the present state of our knowledge we can distinguish the two subgroups Psocoptera and Phthiraptera, although only the Phthiraptera can with confidence be called monophyletic. There is a suspicion that the "Psocoptera" are grouped together on the basis of symplesiomorphy, and that one or another of its subgroups is closely related to the Phthiraptera. Thus if any author needs to present the system of the Psocodea as a whole, no one expects him to present the phylogenetic sister-group relationships more accurately than they are known. If he has reasons for calling the Phthiraptera a suborder of the Psocodea (see p. 186), then he will oppose the "Psocoptera" to this group as a second suborder even though he doubts that the requirements of phylogenetic systematics justify this. He must, however, express his doubt. There is no agreement as to how this should be done; perhaps the simplest method is to place the names of groups not definitely shown to be monophyletic in parentheses.

Such provisional and forced compromises need not be discussed farther here. They do not affect the theory of phylogenetic systematics, or at least only marginally. Actually it should not even be necessary to mention those compromises that explicitly recommend combined, syncretistic systems as the optimal and final aim of systematics. The dangers and finally the nonproductiveness of such systems have been recognized by clear-sighted authors, and were also discussed by us (p. 76). It seems, however, that recommendations of compromise are not always based on a disregard of theoretical considerations. Many authors are misled by considerations of the usability of biological systems in university teaching, or by areas of biology where exact knowledge and presentation of phylogenetic relationships are less important. This probably explains the fact that authors who develop clear ideas of the problems and values of phylogenetic systematics may lose sight of these in the course of one and the same study and so become entangled in obvious contradictions.

Myers (1960), for example, says: "It is my belief . . . that any formal system used by systematists should as closely approximate phylogeny as the limits of materials and human effort permit." But later in the same paper we read that "any classification or system . . . should approximate what is known of relationships, but not to the extent of making the system or classification too complex or esoteric for general biological use." Similarly Stammer (1960) first recognizes that according to the view of "by far the greater number of zoologists interested in systematics . . . the system must be ordered phylogenetically. Only the natural system has true knowledge value for biology; from it new results and new problems constantly arise." But a few pages later he demands that a compilation of species "according to relationships" must not lead to a "disruption of natural, easily recognizable units."

Both of these authors feel that a system in which all recognizable phylogenetic kinship relationships are represented would be too complicated. What is this supposed to mean? If a scientific study aims at determining the phylogenetic kinship relationships in a particular group of animals, it must pursue its aim as far as "the limits of the materials and human effort permit." The results may be presented, as Stammer recommends, in a (phylogenetic tree) "sketch." But phylogenetic tree and "system" are only different ways of presenting the same state of affairs (p. 194). To speak or write in unmistakable terms about what is present in a family tree can be done only in the category of the "phylogenetic system," and consequently a properly drawn phylogenetic tree must be directly translatable into the language of phylogenetic systematics. If it satisfies this requirement, everything has been done to fulfill the theory of phylogenetic systematics. This of course includes taking care that in nomenclatural matters it is not necessary to speak of monophyletic groups in terms such as "group Pterygota plus Thysanura minus Machilidae." But the theory of phylogenetic systematics shows unmistakably how this can be avoided.

An entirely different question is the extent to which all the details presented in special studies (particularly the completeness of the category stages that are distinguished) should be considered in catalogs, textbooks, and handbooks. This depends entirely on the purpose for which such works are meant. Consequently it is needless to worry that, because of the necessity of presenting all the "innumerable ramifications of the phylogenetic tree," in practice there is simply too much demanded of systematics (Stammer 1961). A truly comprehensive catalog, for example (such as the multi-volume *Coleoptorum catalogus*), can take care of this problem, along with the equally troublesome problems of the innumerable difficult questions of nomenclature and synonymy and the delimitation of species.

A last version of our views may perhaps clarify the distinction between phylogenetic and typological systems also for those authors who could so far not recognize it because the typological systems presently in use (which do not permit polyphyletic groups) do reflect the history of the animal world in a particular way, just as the phylogenetic system intends to be the historic presentation of phylogeny. But the requirement that a modern system should "correspond to the phylogenetic development of the group in question" and that "the system should correspond to the phylogeny" (Wagner 1960) is too vague to permit distinction of phylogenetic from typological systematics. There are two ways of describing phylogenetic history: One in which the chronology of the species cleavage processes is set forth as the only supra-individual, real process of phylogeny; and the other in which these same cleavage processes are drawn up according to their epochal significance. Only the first-named way leads to the phylogenetic system; the last-named one leads to the different typological systems, depending on the scale used to evaluate the epochal significance of the events. In Fig. 69 the two ways of writing the history of phylogenesis and the resulting systems are juxtaposed in a concrete example.

If an author or an instructor to whom the differentiation of the structural plans of the organism seems most important sees the epochal events, for example, in the early history of the insects as presented in Fig. 69B, and if he thus presents to his audience a system that reflects *his* valuation of the events (for example, as in Fig. 69B) no one would quarrel with him. He should not be surprised, however, if other authors value the epochal significance of the events differently; nor if events which according to his opinion have had epochal significance in the structural development of the imagines of the insects do not have this significance in the morphological differentiation of the larvae, so that the respective larval and imaginal systems, both of which have been drawn up according to the viewpoints of an epochal presentation of the history of phylogenesis, cannot be brought into congruence.

He must further consider that systems not founded on an exact chronology of the real historical events in phylogeny (but instead are based on a valuation of their differing significance, for example, regarding the structural differentiation of the organisms) are useless for certain investigations into the overall course of phylogeny, because they inevitably include paraphyletic groups in addition to monophyletic ones. Reality, individuality, and, connected with the latter, origin, differentiation, and extinction have however a different significance for monophyletic than for paraphyletic groups. The indiscriminate use of these and other concepts for monophyletic and paraphyletic groups leads as *aequivocatio termi-*

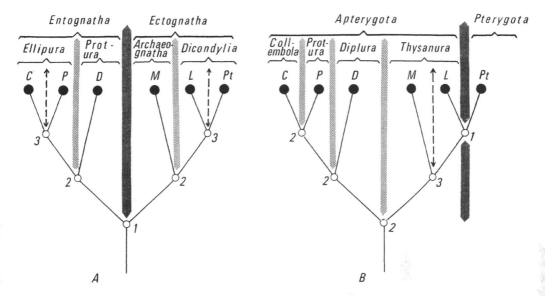

Figure 69. Systematics as the historic presentation of phylogeny. In the phylogenetic system (left) the real-historical events (cleavage processes) of phylogeny are valued according to their chronological sequence (1, 2, 3) and accordingly the relative rank order of the categories is determined. Valuation of the events according to their "epochal significance" (1, 2, 3 in the presentation at right) leads to typological systems.

norum in conclusions by way of the *quaternio terminorum* inevitably to the logical error of *metabasis eis allo genos* and thereby to faulty conclusions.

The attempt to gain insight into the overall course of phylogeny and its conformities to law by an uncritical comparison of paraphyletic and monophyletic groups bears as little hope of achieving valid results as, for example, the attempt to arrive at an understanding of the laws governing the movements of the planets by analysis of planetary movements of which some are described in the concepts of a geocentric, others of a heliocentric universe.

For this reason typological systems, wherever they are used, have only a limited cognitive value, though no one would dispute that they have such. The phylogenetic system, on the other hand, with its exact chronology of all recognizable events (free from value judgments) in the actual history of phylogeny, has the same universal significance for all phylogenetic research, as has, for example, the topographic map as the foundation for all other possible or desirable cartographic presentations in the fields of geography and other geosciences.

The latter statement, then, is the essence of our contention that the phylogenetic system may be regarded, for inherent reasons, as the general reference system of biology.

Bibliography

Abel, O., Paläobiologie und Stammesgeschichte. G. Fischer, Jena 1929.
———, Tiere der Vorzeit in ihrem Lebensraum, in: Das Reich der Tiere, Ergän-
  zungsband. Berlin 1939.
Aczél, M., Morfologia externa y división sistematica de las "Tanypezidiformes" con
  sinopsis de las especies argentinas de "Tylidae" ("Micropezidae") y "Neriidae" (Dipt.)
  Acta zool. lilloana 11, pp. 483-589, 1951.
Alberti, B., Über Dualspezies, Artspaltung und Monophylie. Dtsche. ent. Z. (N. F.)
  2, pp. 211-224, 1955.
———, Wesen und praktische Bedeutung des Gattungs-Begriffes. Bericht 8. Wand.-
  Versamml. dtsch. Ent., pp. 138-147, 1957.
Anderson, E., Supraspecific variation in nature and in classification, from the view-
  point of botany, in: Amer. Nat. 71, pp. 223-235, 1937.
Arldt, T., Die Entwicklung der Kontinente und ihrer Lebewelt I. 2. Aufl., Berlin 1938.
Arnold, R., Das Tier in der Weltgeschichte. Senckenberg-Büch. 8. Frankfurt am Main
  1939.
Bachmetjev, P., Ein Versuch, das periodische System der Schmetterlinge aufzustellen.
  Trav. Soc. Nat. Saratov 4. 3, 1903-04.
Backmann, G., Wachstumszyklen und phylogenetische Entwicklung. Lunds Univ.
  Arsskrift (N.F.) Avd. 2, Bd. 34, Nr. 5, pp. 1-142, 1938.
Barnes, H. F., The biological approach to the species problem in gall midges (Dipt.,
  Cecidomyidae). Ann. ent. fenn. 19, pp. 1-24, 1953.
Bavink, B., Ergebnisse und Probleme der Naturwissenschaften. 7. Aufl., Leipzig 1941.
Beer, G. de, Paedomorphosis. Proc. XVth Intern. Congr. Zool., London, pp. 927-930,
  1959.
Bertalanffy, L. v., Theoretische Biologie I und II. Berlin 1932 und 1942.
———, Die organismische Auffassung und ihre Auswirkungen, in: Der Biologe 10, pp.
  247-258, 337-345, 1941.

Beurlen, K., Vergleichende Stammesgeschichte. Grundlagen, Methoden, Probleme, unter besonderer Berücksichtigung der höheren Krebse, in: Fortschr. Geol. VIII. 26, pp. 317-586, 1930.

————, Die stammesgeschichtlichen Grundlagen der Abstammungslehre. Jena 1937.

Bhatia, M. L., and Keilin, D., On a new case of parasitism of snail (*Vertigo genesii* Gradl.) by a Dipterous larva, in: Parasitology 29, pp. 399-408, 1937.

Bigelow, R. S., Monophyletic classification and evolution. Syst. Zool. 5, pp. 145-146, 1956.

————, Classification and phylogeny. Syst. Zool. 7, pp. 49-59, 1958.

————, Similarity, ancestry and scientific principles. Syst. Zool. 8, pp. 165-168, 1959.

Blackwelder, R. E., The functions and limits of classification. Syst. Zool. 8, pp. 202-211, 1959.

Blagoveschenski, A. V., On the relations between the biochemical properties and the degree of evolutionary development of organisms, in: Biol. gen. 5, pp. 427-500, 1929.

Börner, C., Ueber die Bedeutung der phylogenetischen Betrachtungsweise für die Systematik, in: 3. Wand.Versamml. dtsch. Ent., Giessen 1929, pp. 71-86, 1929.

Boettger, C. R., Die systematische Stellung der Apterygota. Proc. 10th Intern. Congr. Ent. (1956) 1, pp. 509-516, 1958.

Borgmeier, T., Die Wanderameisen der neotropischen Region. Studia ent. 3, pp. 1-717, 1955.

Boyden, A., Homology and analogy. A critical review of the meanings and implications of these concepts in biology. Amer. Midl. Nat. 37, pp. 648-669, 1947.

————, Serology as an aid to systematics. Proc. XVth Intern. Congr. Zool., London, pp. 120-122, 1959.

Brauer, F., Systematisch-Zoologische Studien 1. System und Stammbaum, in: S. B. Akad. Wiss. Wien (math.-nat.) 91, I, pp. 237-272, 1885.

Brown, W. L., Jr., Speciation: The center and the periphery. Proc. 10th Intern. Congr. Ent. (1956) 1, pp. 89-99, 1958.

Bünning, E., Theoretische Grundlagen der Physiologie. Jena 1948.

Cain, A. J., The genus in evolutionary taxonomy. Syst. Zool. 5, pp. 97-109, 1956.

————, The post-Linnean development of taxonomy. Proc. Linn. Soc. Lond. 170, pp. 234-244, 1959.

————, and Harrison, G. A., An analysis of the taxonomist's judgment of affinity. Proc. zool. Soc. Lond. 131, pp. 85-98, 1958.

Callan, H. G., and Spurway, H., A study of meiosis in inter-racial hybrids of the newt Triturus cristatus J. Genet. 50, pp. 235-249, 1951.

Christiansen, K., Geographic variation and the subspecies concept in the Collembolan Entomobryoides guthriei. Syst. Zool. 7, pp. 8-15, 1958.

Clark, R. B., Species and systematics. Syst. Zool. 5, pp. 1-10, 1956.

Claus-Grobben Kühn, Lehrbuch der Zoologie. 10 Aufl., Berlin 1932. (Springer.)

Collier, W. A., Allgemeine Methoden des zoologisch-systematischen Arbeitens; die Nomenklatur, in: Abderhalden, Handbuch der biologischen Arbeitsmethoden Abt. IX, Teil 1, 1. Hälfte, pp. 585-658, 1924.

Cuénot, L., Principes pour l'établissement d'un arbre généalogique du règne animal, in: C.R. Acad. Sci. Paris 209, pp. 736-739, 1939.

————, Remarques sur un essai d'arbre généalogique du règne animal, *ibid.*, 210, pp. 23-27, 1940.

————, Essai d'arbre généalogique du règne animal, *ibid.*, pp. 196-199, 1940.

Cushman, J. A., Foraminifera, their classification and economic use. 2. Aufl., Sharon 1933.

Dabelow, A., Abschnitt Vergleichende Morphologie der Wirbeltiere, in: Fortschr. Zool. (N.F.) 6, pp. 1-70, Jena 1942.

Dacqué, E., Organische Morphologie und Paläontologie. Berlin 1935.
Danser, B. H., A theory of systematics. Bibliogr. biotheor. IV, 3, pp. 117-180. Leiden 1950.
Darlington, P. J., Jr., The geographical distribution of cold-blooded vertebrates. Quart. Rev. Biol. 23, pp. 1-26, 105-123, 1948.
――――, Zoogeography. The geographical distribution of animals. New York 1957.
Dessauer, H. C., and Fox, W., Characteristic electrophoretic patterns of plasma proteins of orders of Amphibia and Reptilia. Science 124, pp. 225-226, 1956.
Diels, L., Die Methoden der Phytographie und der Systematik der Pflanzen, in: Abderhalden, Handbuch der biologischen Arbeitsmethoden, Abt. XI, Teil 1, Heft 2, pp. 67-190, 1921.
Dingler, H., Die philosophische Begründung der Deszendenztheorie, in: Heberer, Die Evolution der Organismen, pp. 3-19, Jena 1943.
Dingler, M., Ueber den Einfluss des Lebensraumes auf die Artenbildung, in: 3. Wand.Versamml. dtsch. Ent., Giessen, pp. 31-47, 1929.
Dobzhansky, T. Die genetischen Grundlagen der Artbildung. Deutsch von W. Lerche. Jena 1939.
――――, Microgeographic variation in Drosophila pseudoobscura, in: Proc. nat. Acad. Sci. U.S.A. 25, pp. 311-314, 1939a.
――――, and Streisinger, G., Experiments on sexual isolation in Drosophila. II. Proc. nat. Acad. Sci. U.S.A. 30, pp. 340-345, 1944.
Döderlein, L., Ueber die Beziehungen naheverwandter Formen zueinander, in: Z. Morph. Anthr. 4, pp. 394-442, 1902.
Dollo, L., Les lois de l'évolution, in: Bull. Soc. belge Géol. Pal. Hydr. 7, pp. 164-166, 1893.
Dunn, E. R., The herpetological fauna of the Americas. Copeia 1931, pp. 106-119, 1931.
Durham, J. W., and Melville, R. V., A classification of Echinoids. J. Paleont. 31, pp. 242-272, 1957.
Edwards, J. G., A new approach to infraspecific categories. Syst. Zool. 3, pp. 1-20, 1954.
Ehrlich, P. R., Problems of higher classification. Syst. Zool. 7, pp. 180-184, 1958.
Eichler, W., Topographische Spezialisation bei Ektoparasiten. Z. Parasitenk. 11, pp. 205-214, 1940.
――――, Wirtsspezifität und stammesgeschichtliche Gleichläufigkeit (Fahrenholzsche Regel) bei Parasiten im allgemeinen und bei Mallophagen im besonderen, in: Zool. Anz. 132, pp. 254-262, 1941.
――――, Die Entfaltungsregel und andere Gesetzmässigkeiten in den parasitogenetischen Beziehungen der Mallophagen und anderer ständiger Parasiten zu ihren Wirten, in: Zool. Anz. 137, pp. 77-83, 1942.
――――, Evolutionsfragen und Wirtsspezifität. Biol. Zbl. 67, pp. 373-406, 1948.
――――, Stammesgeschichtliche Parallelbeziehungen im Wirt- Parasit- Verhältnis von Fischparasiten. Z. Fisch. (N.F.) 1, pp. 301-308, 1953.
Eisler, R., Wörterbuch der philosophischen Begriffe. 4. Aufl., Berlin (3. Band 1930).
Ekman, S., Tiergeographie des Meeres. Leipzig 1935.
Emden, F. I. van, The taxonomic significance of the characters of immature insects. Annu. Rev. Ent. 2, pp. 91-106, 1957.
――――, The two larval forms of Meloe violaceus Marsh., and species distinguishable only in the early stages. Proc. 10th Intern. Congr. Ent. (1956) 1, pp. 217-221, 1958.
――――, Evolution of Tachinidae and their parasitism. Proc. XVth Intern. Congr. Zool., London (1958), pp. 664-666, 1959.
Erdmann, Rhoda, Verwandtschaftsbeziehungen der Anurenfamilien, geprüft durch

Transplantationsversuche gezüchteter Haut, in: Roux's Arch. Entwicklungsmechanik 112, pp. 739-806, 1927.

Faegri, K., The species problem, in: Nature 136, pp. 954-955, 1936.

Fechner, G. T., Einige Ideen zur Schöpfungs- und Entwicklungsgeschichte der Organismen. Leipzig 1873.

Franz, V., Der biologische Fortschritt. Die Theorie der organismengeschichtlichen Vervollkommnung. Jena 1935.

Friederichs, K., Grundsätzliches über die Lebenseinheiten höherer Ordnung und den ökologischen Einheitsfaktor, in: Naturwissenschaften 15, pp. 153-157, 182-186, 1927.

——, Die Grundfragen und Gesetzmässigkeiten der land- und forstwirtschaftlichen Zoologie, Band I und II. Berlin 1930.

——, Oekologie, in: Bios 7, Lpz. 1937.

——, Zur ökologischen Terminologie, in: Anz. Schädlingsk. 18, pp. 97-98, 1942.

Fuhrmann, O., Die Cestoden der Vögel, in: Zool. Jb., Suppl. 10, pp. 1-232, 1908.

Gehlen, A., Der Mensch 1940. Berlin 1940.

——, Zur Systematik der Anthropologie, in: N. Hartmann, Systematische Philosophie, pp. 1-54, 1942.

Gregg, J. R., The language of taxonomy. An application of symbolic logic to the study of classificatory systems. New York 1954.

Günther, K., Systematik und Stammesgeschichte der Tiere 1939-1953. Fortschr. Zool. (N.F.) 10, pp. 33-278, 1956.

——, Systematik und Stammesgeschichte der Tiere 1954-1959. Fortschr. Zool. (N.F.) 14, pp. 269-547, 1962.

Haas, O., Besprechung von Anderson, E., Plants, Man and Life. Syst. Zool. 3, p. 65, 1954.

Haase-Bessell, G., Der Evolutionsgedanke in seiner heutigen Fassung. Jena 1941.

Hagmeier, E. M. Inapplicability of the subspecies concept to North American marten. Syst. Zool. 7, pp. 1-7, 1958.

Handlirsch, A., Kapitel: Die systematischen Grundbegriffe und Phylogenie und Stammesgeschichte in: Schröder, C., Handbuch der Entomologie, Band III, Jena 1925.

Hargis, W. J., Jr., A suggestion for the standardization of the higher systematic categories. Syst. Zool. 5, pp. 42-46, 1956.

Harrison, L., Host and parasite. Proc. Linn. Soc. N.S.W. 35, pp. IX-XXXI, 1928.

Hartmann, Max., Allgemeine Biologie. Ihre Aufgaben, ihr gegenwärtiger Stand und ihre Methode, in: Biologie 1, pp. 239-248, 1932.

——, Die Sexualität. Jena 1943.

——, Allgemeine Biologie. 3. Aufl. Jena 1947.

Hartmann, Nicolai, Systematische Philosophie. Stuttgart u. Berlin 1942.

Heberer, G., Die Evolution der Organismen. Jena 1943.

——, Experimentelle Phylogenetik and Typensprunglehre, in: Der Biologe 12, pp. 248-253, 1943a.

——, Begriff und Bedeutung der parallelen Evolution. Verh. dtsch. zool. Ges. 1952, pp. 435-442, 1953.

——, Zum Problem der additiven Typogenese, in: Hedberg, O., Systematics of today, pp. 40-47, Uppsala and Wiesbaden 1958.

Heintz, A., Der Stammbusch des Tierreiches, in: Natur u. Volk 69, pp. 524-534, 1939.

Helmcke, J. G., Brachiopoda, in: Kükenthal-Krumbach, Handbuch der Zoologie, 3. Band, 2. Hälfte, pp. (5) 139-(5) 262, 1939.

Henbest, L. G., Significance of evolutionary explosions for diastrophic division of earth history. J. Paleont. 26, pp. 299-318, 1952.

Hendel, F., Zweiflügler oder Diptera II: Allgemeiner Teil, in: Dahl, die Tierwelt Deutschlands und der angrenzenden Meeresteile. Teil 11. Jena 1928.

Hennig, E., Bahnen des Lebensstromes, in: Aus der Heimat 40, pp. 234-237, 1927.
——, Wesen und Wege der Paläontologie. Berlin 1932.
Hennig, W., Revision der Gattung *Draco*, in: Temminckia 1, pp. 153-220, 1936.
——, Ueber einige Gesetzmässigkeiten der geographischen Variation in der Reptilien-gattung *Draco* L.: parallele und konvergente Rassenbildung, in: Biol. Zbl. 56, pp. 549-559, 1936a.
——, Die Gattung *Rachicerus* Walker und ihre Verwandten im Baltischen Bernstein, in: Zool. Anz. 123, pp. 33-41, 1938.
——, Kritische Bemerkungen zum phylogenetischen System der Insekten. Beitr. Ent. 3, Sonderheft, pp. 1-85, 1953.
——, Flügelgeäder und System der Dipteren, unter Berücksichtigung der aus dem Mesozoikum beschriebenen Formen Beitr. Ent., 4, pp. 245-388, 1954.
——, Systematik und Phylogenese. Ber. Hundertj. dtsch. ent. Ges. (1956), pp. 50-71, 1957.
——, Die Familien der Diptera Schizophora und ihre phylogenetischen Ver-wandtschaftsbeziehungen. Beitr. Ent. 8, pp. 508-688, 1958.
——, Die Dipteren-Fauna von Neuseeland als systematisches und tiergeographisches Problem. Beitr. Ent. 10, pp. 221-329, 1960.
Hering, M., Die Oekologie der blattminierenden Insektenlarven, in: Zool. Baust. 1, 2. Berlin 1926.
——, Dualspecies und Unterart-Entstehung. *Ceriocera ceratocera* (Hend.) u. sub-spec. *microceras* nov., in: Dtsch. ent. Z. 1935, pp. 207-210, 1935.
——, Subspecies in statu nascendi, in: Zool. Anz. 114, pp. 266-271, 1936.
——, Die Nahrungswahl phytophager Insekten. Verh. dtsch. Ges. angew. Ent. (13) (1954), pp. 29-38, 1955.
——, Gedanken zum Synchronismus der Entwicklung phytophager Insekten und ihrer Futterpflanzen. Scient. Nat. Studia et opuscula in honorem septuagenarii W. Weisbach civis Hagae comitum ab amicis collegisque conscripta. A. D. MCMLIX edita ab Ingeborg Feitz-Weisbach, pp. 7-31, Den Haag 1960.
Herting, B., Artbildung auf dem Wege über ökologische Rassen bei parasitischen Fliegen (Dipt., Tachinidae). Verh. dtsch. zool. Ges. Münster/Westf. 1959, pp. 171-174, 1960.
Hertwig, P., Artbastarde bei Tieren, in: Handbuch der Vererbungswissenschaft II B, Berlin 1936.
Hertwig, R., Die Abstammungslehre, in: Die Kultur der Gegenwart, Abt. 4, 4, Teil 3, pp. 1-91, 1914.
Heyden, C. v., Ueber ein sonderbar gestaltetes Thierchen, in: Okens Isis 1823, pp. 1247-49, und Nachtrag, *ibid.*, 1825, pp. 588-589.
Horn, W., Heteropod-Zoology and Entomological Complexes, in: Ent. News 39, pp. 172-178, 1928.
——, On the splitting influence of the increase of entomological knowledge and on the enigma of species, in: 4th Intern. Congr. Ent., Ithaka (1928) II, pp. 500-507, 1929.
——, Ueber die Zukunft der Insekten-Systematik, in: Anz. Schädlingsk. 5, pp. 40-45, 1929a.
Huene, F. v., Die stammesgeschichtliche Gestalt der Wirbeltiere—ein Lebensablauf, in: Paläont. Z. 22, pp. 55-62, 1940.
Huff, C. G., Studies on the evolution of some disease-producing organisms, in: Quart. Rev. Biol. 13, pp. 196-206, 1938.
Humphries, C., Neue *Trichocladius*-Arten, in: Stettin. ent. Ztg. 98, pp. 185-195, 1937.
Huxley, J. S., Clines: An auxiliary method in taxonomy. Bijdr. Dierk. 27, pp. 491-520, 1939.

————, Towards the new systematics. The new systematics, pp. 1-46, London 1940.

————, Evolutionary processes and taxonomy, in: Hedberg, O., Systematics of today, pp. 21-39, Uppsala and Wiesbaden 1958.

Ihle, I. E. W., Tunicata, in: Kükenthal-Krumbach, Handbuch der Zoologie, Vol. 5, 2. Hälfte, p. 525, 1935.

Jeannel, R., Origin et évolution des Insectes. Verh. 8. intern. Kongr. Ent., Stockholm, pp. 80-86, 1950.

Johnson, M. L., and Wicks, M. J., Serum protein electrophoresis in mammals. Taxonomic implications. Syst. Zool. 8, pp. 88-95, 1959.

Karny, H. H., Die Methoden der phylogenetischen (stammesgeschichtlichen) Forschung, in: Abderhalden, Handbuch der biologischen Arbeitsmethoden Abt. IX, Teil 3, pp. 211-500, 1925.

Kempermann, C. T., Am Wendepunkt der Stammesgeschichte. Jena 1936.

Kessel, E. L., Sex limited polychromatisme in *Lasiophthicus pyrastri* L., in: Pan-Pacif. Ent. 2, p. 159, 1926.

————, The mating activities of balloon flies. Syst. Zool. 4, pp. 97-104, 1955.

Kienle, H., Die Masstäbe des Kosmos. Vorträge und Schriften, Dtsch. Akad. wiss. Berl. 24, pp. 1-29, 1948.

Kinsey, A. C., The origin of higher categories in *Cynips*, in: Indiana Univ. Publ., Sci. Ser. no. 4, pp. 1-334, 1936.

————, An evolutionary analysis of insular and continental species, in: Proc. nat. Acad. Sci. U.S.A. 23, pp. 5-11, 1936a.

————, Super-specific variation in nature and in classification, in: Amer. Nat. 71, pp. 206-222, 1937.

Kiriakoff, S. J., De huidige Problemen van de Taxonomische Terminologie in de Dierkunde. Brussels 1948.

————, Das Vavilov'sche Gesetz, die Taxonomie and die Zoogeographie. Zool. Anz. 156, pp. 277-284, 1956.

Kleinschmidt, O., Die Formenkreislehre und das Weltwerden des Lebens. Halle 1926.

————, Der grösste Skandal in der Geschichte der Ornithologie, in: Falco 36, pp. 33-48, 1940.

Klingstedt, H., A taxonomic survey of the genus *Cyrnus* Steph. including the description of a new species, with some remarks on the principles of taxonomy, in: Acta Soc. Fauna Flora fenn. 60, 1937.

Krogh, C. V., Serologische Verwandtschaft oder stammesgeschichtliche Verwandtschaft?, in: Zool. Anz. 123, pp. 206-213, 1938.

Krüger, F., Tanytarsus-Studien I. Die Subsectio *Atanytarsus*. Zugleich variationsstatistische Untersuchungen zum Problem der Artbildung bei Chironomiden, in: Arch. Hydrobiol. 33, pp. 208-256, 1938.

Krüger, P., Abstammung und Biochemie, in: Biol. gen. 6, pp. 483-510, 1930.

Kühnelt, W., Prinzipien der Systematik, in: Bertalanffy, Handbuch der Biologie VI, 1, pp. 1-16, 1942.

Kükenthal, W., Die Bedeutung der Verbreitung mariner Bodentiere für die Paläogeographie, in: S. B. Ges. naturf. Fr. Berl., 1919, pp. 208-218.

Kutscher, F., and Ackermann, D., Vergleichend-physiologische Untersuchung von Extrakten verschiedener Tierklassen auf tierische Alkaloide, eine Zusammenfassung, in: Z. Biol. 84, pp. 181-192, 1926.

Lam, H. J., Phylogenetic symbols, past and present, in: Acta biotheor. A 2, pp. 153-194, 1936.

Lathrop, F. H., and Nickels, C. B., The blueberry maggot from an ecological viewpoint. Ann. ent. Soc. Amer. 24, pp. 260-281, 1931.

Linder, E., Ueber einige bemerkenswerte Konvergenzen im System der Stratiomyiiden, in: Verh. 7. intern. Kongr. Ent. 1, pp. 236-239, 1939.

Mägdefrau, K., Paläobiologie der Pflanzen. Jena 1942.

Mansfeld, R., Ziele und Wege der botanischen Systematik. Wiss. Ann. 1, pp. 329-345, 1952.

Manwell, R. D., Intraspezific variation in parasitic protozoa. Syst. Zool. 6, pp. 2-6, 1957.

Marinelli, W. v., Zoologie und Abstammungslehre, in: Palaeobiologica 7, pp. 169-196, 1939.

Martini, E., Diskussionsbemerkungen zu den Tagesthemata "Artbegriff" und "Phylogenie," in: 3. Wand.Versamml. dtsch. Ent., Giessen, 1929, pp. 94-98.

———, Zur Frage der biologischen Arten. Ein Beitrag zur Erörterung über die derzeitige Lage der systematischen und sammelnden Zoologie, in: Arb. Physiol. angew. Ent. Berl. 5, pp. 33-43, 1938.

Maslin, T. P., Morphological criteria of phyletic relationships. Syst. Zool. 1, pp. 49-70, 1952.

May, E., Kleiner Grundriss der Naturphilosophie. Meisenheim/Glan 1949.

Mayr, E., Systematics and the origin of species from the viewpoint of a zoologist. New York 1942.

———, Die denkmöglichen Formen der Artentstehung. Rev. suisse Zool. 64, pp. 219-235, 1957.

———, The evolutionary significance of the systematic categories, in: Hedberg, O., Systematics of today, pp. 13-20, Uppsala and Wiesbaden 1958.

———, Linsley, G., and Usinger, R. L., Methods and principles of systematic zoology. New York, Toronto, London 1953.

McGuire, J. N., Jr., and Wirth, W. W., The discriminant function in taxonomic research. Proc. 10th Intern. Congr. Ent. (1956) 1, pp. 387-393, 1958.

Meglitsch, P. A., On the nature of the species. Syst. Zool. 3, pp. 49-65, 1954.

Meyrick, E., On the classification of the Australian Pyralidina, in: Trans. R. ent. Soc. Lond. 1884, p. 277.

Mez, C., Serum-Reaktionen zur Feststellung von Verwandtschaftsverhältnissen im Pflanzenreich, in: Abderhalden, Handbuch der biologischen Arbeitsmethoden Abt. IX, Teil 1, pp. 1059-94, 1924.

———, Theorien der Stammesgeschichte, in: Schriften Königsberger gelehrten Gesellschaft, Naturw. 3. Klasse, 4, pp. 99-128, 1926.

———, and Ziegenspeck, H., Der Königsberger serodiagnostische Stammbaum, in: Bot. Arch. 13, p. 483, 1926.

Michaelis, L., Einführung in die Mathematik für Biologen und Chemiker. 3. Aufl., Berlin 1927.

Michaelsen, W., Das Wesen der Systematik, den jungen Kollegen an dem Beispiel des modernen Oligochätensystems erläutert, in: Zool. Anz. 109, pp. 1-19, 1935.

Michener, C. D., Life-history studies in insect systematics. Syst. Zool. 2, pp. 112-118, 1953.

———, Some bases for higher categories in classification. Syst. Zool. 6, pp. 160-173, 1957.

———, and Sokal, R. R., A quantitive approach to a problem in classification. Evolution 11, pp. 130-162, 1952.

Möbius, K., Die Bildung, Geltung und Bezeichnung der Artbegriffe, in: Zool. Jb. 1, pp. 1-36, 1886.

Mollison, T., Serodiagnostik als Methode der Tiersystematik und Anthropologie, in: Abderhalden, Handbuch der biologischen Arbeitsmethoden Abt. IX, Teil 1, 1. Hälfte, pp. 553-584, 1924.

Moore, J. A., Patterns of evolution in the genus Rana, in: Jepsen, Simpson, and Mayr, Genetics, paleontology and evolution, pp. 315-338, Princeton 1949.

Mühlmann, W. E., Geschichtliche Bedingungen, Methoden und Aufgaben der Völkerkunde, in: Preuss-Thurnwald, Lehrbuch der Völkerkunde, pp. 1-43. 2. Aufl., Stuttgart 1939.

Müller, A. H., Der Grossablauf der stammesgeschichtlichen Entwicklung. Jena 1955.

Müller, H. J., Reversibility in evolution considered from the standpoint of genetics, in: Biol. Rev. 14, pp. 167-209, 1939.

Müller-Freienfels, R., in: Eislers Handwörterbuch der Philosophie. 2. Aufl., Berlin 1922.

Myers, G. S., Some reflections on phylogenetic and typological taxonomy. Syst. Zool. 9, pp. 37-41, 1960.

Naef, A., Idealistische Morphologie und Phylogenetik. Jena 1919.

———, Ueber Morphologie und Stammesgeschichte, in: Vjschr. naturf. Ges. Zürich 70, pp. 234-240, 1925.

———, Phylogenie der Tiere, in: Baur and Hartmann, Handbuch der Vererbungswissenschaft III, 1 (Liefg. 13), 1931.

Nägeli, C., Entstehung und Begriff der naturhistorischen Art. 2. Aufl., 1865.

Newell, N. D., Periodicity in invertebrate evolution. J. Paleont. 26, pp. 371-385, 1952.

O'Rourke, F. J., Serological tools in entomological research. Ent. Gaz. 9, pp. 63-72, 1958.

Paramonow, S. J., Die Methoden der modernen Zoosystematik (Zoogeographie), in: Trav. Mus. zool. Kiew 14, pp. 3-12, 1935.

———, Gegenwärtige Systematik, ihre Methoden und Aufgaben, ibid., 13, pp. 3-23, 1935a.

———, Das Problem der Artbildung und das Areal, ibid., 15, pp. 24-26, 1935b.

———, Ob das System der Tiere phylogenetisch sein soll?, ibid., 19, pp. 197-212, 1937.

———, Modern zoological taxonomy, its theoretical and practical problems, in: Zool. J. 18, pp. 7-26, 1939.

———, Ein neues System der niederen taxonomischen Einheiten in Form einer Bestimmungstabelle, in: Arb. morph. taxon. Ent. Berl. 11, pp. 33-40, 1944.

Pickett, A. D., and Neary, M. E., Further studies in Rhagoletis pomonella (Walsh). Sci. Agric. 20, pp. 551-553, Ottawa 1940.

Pictet, A., Sur les croisements des races géographiques de Lépidoptères de pays très éloignés, in: Mitt. schweiz. ent. Ges. 16, pp. 706-715, 1936.

Plate, L., Prinzipien der Systematik mit besonderer Berücksichtigung des Systems der Tiere, in: Kultur der Gegenwart, Abt. 4, 4, Teil 3, pp. 92-164, 1914.

———, Ueber Vervollkommnung, Anpassung und Unterscheidung von niederen und höheren Tieren, in: Zool. Jb. 45 (Festschrift Hesse), pp. 745-798, 1928.

Poche, F., Zur Vereinheitlichung der Bezeichnung und exakteren Verwendung der systematischen Kategorien und zur rationelleren Benennung der supragenerischen Gruppen. Verh. 8. intern. Zool. Kongr. Graz, 1910, pp. 819-850, 1950.

Prell, H., Diskussionsbemerkungen zum Tagesthema "Artbegriff," in: 3. Wand. Versamml. dtsch. Ent., Giessen, pp. 62-65, 1929.

Przibram, H., Phylogenese. Eine Zusammenfassung der durch Versuche ermittelten Gesetzmässigkeit tierischer Artbildung (Arteigenheit, Artübertragung, Artwandlung), in: Experimentalzoologie 3, Leipzig und Wien 1910.

Remane, A., Kinorrhyncha, in: Kükenthal-Krumbach, Handbuch der Zoologie, 2. Band, 1. Hälfte, pp. (4)187-(4)248, 1929.

———, Die Grundlagen des natürlichen Systems, der vergleichenden Anatomie und der Phylogenetik. Theoretische Morphologie und Systematik I. Leipzig 1952.

Rensch, B., Das Prinzip geographischer Rassenkreise und das Problem der Artbildung. Berlin 1929.

————, Zoologische Systematik und Artbildungsproblem, in: Verh. dtsch. zool. Ges., pp. 19-83, 1933.

————, Kurze Anweisung für zoologisch-systematische Studien. Leipzig 1934.

————, Typen der Artbildung, in: Biol. Rev. 14, pp. 180-222, 1939.

Rogers, J. S., Symposium: Subspecies and clines. Syst. Zool. 3, pp. 97-126, 133, 1954.

Romer, A. S., Vertebrate paleontology. 4th Impr. Chicago 1950.

————, Explosive evolution. Zool. Jb., Abt. Syst., 88, pp. 79-90, 1960.

Rosa, D., La riduzione progressiva della variabilità, Turin 1899, deutsch von H. Bossard, Die progressive Reduktion der Variabilität und ihre Beziehungen zum Aussterben und zur Entstehung der Arten. Jena 1903.

Rosenfeld, L., and Goldmann, S., Daten über die Untersuchung von Oxydationsprozessen vom phylogenetischen Gesichtspunkte aus, in: Biochem. Z. (1938) 12, pp. 369-383, 1939.

Ross, H. H., The evolution of the insect orders. Ent. News 66, pp. 197-208, 1955.

Roux, W., Prinzipielle Sonderung von Naturgesetz und Regel, vom Wirken und Vorkommen, in: S. B. preuss. Akad. Wiss. 1920, pp. 525-554, 1920.

Rozeboom, L. E., and Kitzmiller, J. B., Hybridization and speciation in mosquitoes. Annu. Rev. Ent. 3, pp. 231-248, 1958.

Rubzov, J. A. K., Evoliuzii krowososutschtich moschek (Simuliidae, Dipt.), in: Bull. Akad. Sci. URSS. (Cl. Sci. math.-nat.), pp. 1289-1327, 1937.

————, On the evolution of bot-flies (Gastrophilidae) in connection with the history of their hosts, in: Zool. J. 18, pp. 669-684, 1939.

Rühm, W., Spiegeln die ipidenspezifischen Nematoden die Verwandtschaft ihrer Wirte wieder? 7. Wand.Versamml. dtsch. Ent. (1954), pp. 81-90, 1955.

Rüschkamp, F., Systematik und Stammegeschichte, in: Ent. Mitt. 16, pp. 420-422, 1927.

Sachtleben, H., Zur Priorität des Satzes von der Irreversibilität der Entwicklung. Beitr. Ent. 1, p. 93, 1951.

Schilder, F. A., Lehrbuch der Allgemeinen Zoogeographie. Jena 1956.

Schindewolf, O. H., Prinzipienfragen der biologischen Systematik, in: Palaeont. Z. 9, pp. 122-169, 1927.

————, Paläontologie, Entwicklungslehre und Genetik. Kritik und Synthese. Bornträger, Berlin 1936.

————, Beobachtungen und Gedanken zur Deszendenzlehre, in: Acta biotheor. A 3, pp. 195-212, 1937.

————, Zur Theorie der Artbildung, in: S. B. Ges. naturf. Fr. Berl. 1939, pp. 368-384, 1940.

————, Entwicklung im Lichte der Paläontologie, in: Der Biologe 11, pp. 113-125, 1942.

————, Zur Frage der sprunghaften Entwicklung, ibid., 12, pp. 238-247, 1943.

Schmalfuss, H., and Werner, H., Chemismus der Entstehung von Eigenschaften, in: Z. indukt. Abstamm. u. Vererb.Lehre 41, pp. 285-358, 1926.

Schmitz, H., Nepenthes-Phoriden, in: Arch. Hydrobiol., Suppl. IX, pp. 449-471, 1931.

Schwanitz, F., Ein Kreuzzug gegen die Abstammungslehre, in: Der Biologe 9, pp. 407-413, 1940.

————, Genetik und Evolutionsforschung bei Pflanzen, in: Heberer, G., Die Evolution der Organismen, pp. 430-478, 1943.

Séguy, E., Muscidae, in: Genera Insect., Fasc. 205, 1937.

Sewertzoff, A. N., Morphologische Gesetzmässigkeiten der Evolution. Jena 1931.

Simpson, G. G., Supra-specific variation in Nature and in Classification from the viewpoint of Paleontology, in: Amer. Nat. 71, pp. 236-267, 1937.

————, Turtles and the origin of the fauna of Latin America. Amer. J. Sci. 241, pp. 413-429, 1943.

————, Tempo and mode in evolution. New York 1944.

————, The principles of classification and the classification of mammals. Bull. Amer. Mus. nat. Hist. 85, XV + 350 pp., 1945.

————, Essay-Review of recent works on evolutionary theory by Rensch, Zimmermann, and Schindewolf. Evolution 3, pp. 178-184, 1949.

————, The species concept. Evolution 5, pp. 285-298, 1951.

————, Evolution and geography. An essay on historical biogeography, with special reference to mammals. Eugene, Oregon 1953.

————, The nature and origin of supraspecific taxa. Cold Spr. Harb. Symp. quant. Biol. 24, pp. 255-271, 1959.

————, Mesozoic mammals and the polyphyletic origin of mammals. Evolution 13, pp. 405-414, 1959.

Simroth, H., Die Aufklärung der südafrikanischen Nacktschneckenfauna, auf Grund des von Herrn Dr. L. Schultze mitgebrachten Materials, in: Zool. Anz. 31, pp. 792-798, 1907.

Smirnov, E., Ueber den Bau der systematischen Kategorien, in: Rev. zool. russe 3, pp. 358-391, 1923.

————, Probleme der exakten Systematik und Wege zu ihrer Lösung, in: Zool. Anz. 61, pp. 1-14, 1924.

————, The theory of type and the natural system, in: Z. indukt. Abstamm. u. Vererb.-Lehre 37, pp. 28-66, 1925.

————, Ueber die Phylogenese der Kongregationen, in: Biol. gen. 2, pp. 241-257, 1926.

————, Mathematische Studien über individuelle und Kongregationsvariabilität, in: Verh. V. intern. Kongr. Vererbungswissensch. Berlin 1927 (Z. indukt. Abstamm. u. Vererb.Lehre, Suppl. II), pp. 1373-92, 1928.

————, Der Begriff der Art vom taxonomischen Standpunkte aus gesehen, in: Zool. J. Moskau 17, pp. 387-418, 1938.

Söderström, A., Ueber evolutionistische Divergenzmorphologie und idealistische "phylogenetische" Morphologie, 48 pp., Uppsala 1927.

Sokal, R. R., Quantification of systematic relationships and of phylogenetic trends. Proc. 10th Intern. Congr. Ent. (1956) 1, pp. 409-415, 1958.

————, and Michener, C. D., A statistical method for evaluating systematic relationships. Kans. Univ. Sci. Bull. 38, pp. 1409-38, 1958.

Spann, O., Erkenne Dich selbst. Jena 1935.

Spix, von, Ueber eine neue Landschnecken-Gattung (Scutelligera Ammerlandia) in Ammerland am Starnberger See in Bayern gefunden, in: Denkschr. K. bayer. Akad. Wiss. 9, pp. 121-124, 1825.

Stammer, H. J., Oekologische Wechselbeziehungen zwischen Insekten und anderen Tiergruppen. 7. Wand.Versamml. dtsch. Ent. (1954), pp. 12-61, 1955.

————, Gedanken zu den parasitophyletischen Regeln und zur Evolution der Parasiten. Zool. Anz. 159, pp. 255-267, 1957.

————, "Trends" in der Phylogenie der Tiere; Ektogenese und Autogenese. Zool. Anz. 162, pp. 187-208, 1959.

————, Neue Wege der Insektensystematik. Verh. Intern. Kongr. Ent. Wien, Vol. 1, pp. 1-7, 1960.

Steiner, B., Stilgesetzliche Morphologie. 1937.

Stephen, W. P., Haemolymph proteins and their use in taxonomic studies. Proc. 10th Intern. Congr. Ent. (1956) 1, pp. 395-400, 1958.

Stresemann, E., Die Entwicklung der Ornithologie. Berlin 1951.

Stunkard, H. W., Intraspecific variation in parasitic flatworms. Syst. Zool. 6, pp. 7-18, 1957.

250   PHYLOGENETIC SYSTEMATICS

Szidat, L., Beiträge zum Aufbau eines natürlichen Systems der Trematoden I., in: Z. Parasitenk. 11, pp. 239-283, 1940.
———, Geschichte, Anwendung und einige Folgerungen aus den parasitogenetischen Regeln. Z. Parasitenk. 17, pp. 237-268, 1956.
Thienemann, A., Unser Bild der lebenden Natur, in: 90./91. Jber. naturh. Ges. Hannover, pp. 27-51, 1940.
Thompson, W. R., The philosophical foundations of systematics. Canad. Ent. 84, pp. 1-16, 1952.
———, Systematics: The ideal and the reality. Boll. Lab. Zool. Portici 33, pp. 320-329, 1956.
Thorpe, W. H., Further studies on pre-imaginal olfactory conditioning in insects, in: Proc. roy. Soc. Lond. (B) 127, pp. 424-433, 1939.
Tillyard, R. J., The panorpoid complex. Part 3. The wing venation. Proc. Linn. Soc. N.S.W. 44, pp. 533-718, 1919.
Torrey, T. W., Organisms on time, in: Quart. Rev. Biol. 14, pp. 275-288, 1939.
Troll, W., Phylogenetische oder idealistische Morphologie? Eine Berichtigung, in: Botan. Arch. 40, 1940.
———, Biomorphologie und Biosystematik als typologische Wissenschaften. Studium gen. 4, pp. 376-389, 1951.
Tschulok, S., Das System der Biologie in Forschung und Lehre. Jena 1910.
———, Deszendenzlehre. Jena 1922.
Tuxen, S. L., Relationships of Protura. Proc. 10th Intern. Congr. Ent. (1956) 1, pp. 493-497, 1958.
Uhlmann, E., Entwicklungsgedanke und Artbegriff. Jena 1923.
Ulrich, H., Kapitel Allgemeine Genetik (einschliesslich Genphysiologie), Genetik und Artbildung, in: Fortsch. Zool. (N. F.) 6, pp. 215-267, 1942.
Vanzolini, P. E., and Guimarães, L. R., South American land mammals and their lice. Evolution 9, pp. 345-347, 1955.
Vavilov, N. L., The law of homologous series variation, in: J. Genet. 12, pp. 47-89, 1922.
Wahlert, G. v., Weitere Untersuchungen zur Phylogenie der Schwanzlurche. Formenzahl und Differenzierungsgrad als umweltabhängige Eigenschaften höherer Kategorien. Zool. Anz., Suppl. 20 (Verh. dtsch. Zool. Ges. 1956), pp. 347-352, 1957.
Weidmann, U., Über den systematischen Wert von Balzhandlungen bei Drosophila. Rev. suisse Zool. 58, pp. 502-511, 1951.
Wein, H., Das Problem des Relativismus, in: N. Hartmann, Systematische Philosophie, pp. 431-560, 1942.
Wenzl, A., Metaphysik der Biologie von Heute. Leipzig 1938.
Wettstein, R., Die geographisch-morphologische Methode in der Botanik. Wien 1899.
Wettstein, O., Systematik und Stammesgeschichte der Wirbeltiere, in: Fortschr. Zool. (N. F.) 5, pp. 97-134, 1941 und 6, pp. 98-115, 1942.
———, Systematik und Stammesgeschichte der Wirbeltiere. Fortschr. Zool. (N. F.) 1, pp. 203-225, 1957.
White, M. J. D., Cytological evidence on the phylogeny and classification of the Diptera. Evolution 3, pp. 252-261, 1949.
Wilckens, O., Stammgarben, in: Z. induckt. Abstamm. u. Vererb.Lehre 20, pp. 241-261, 1919.
Wilson, H. F., and Doner, M. H., The historical development of insect classification. New York 1937.
Woltereck, R., Grundzüge einer allgemeinen Biologie. 2. Aufl., Stuttgart 1940.
Woodger, J. H., From biology to mathematics. Brit. J. Phil. Sci. 3, pp. 1-21, 1952.

Zarapkin, S. R., Das Divergenzprinzip in der Bestimmung kleiner systematischer Kategorien, in: Verh. 7. intern. Kongr. Ent. 1, pp. 494-518, 1939.

Ziehen, T., Erkenntnistheorie, 2. Aufl. I and II, Jena 1934 and 1939.

Zimmermann, W., Arbeitsweise der botanischen Phylogenetik und anderer Gruppierungswissenschaften, in: Abderhalden, Handbuch der biologischen Arbeitsmethoden Abt. 3, 2, Teil 9, pp. 941-1053, 1931 (1937).

——, Die Methoden der Phylogenetik, in: Heberer, G., Die Evolution der Organismen, pp. 20-56, Jena 1943.

——, Evolution. Geschichte ihrer Probleme und Erkenntnisse. Freiburg und München 1953.

Zündorf, W., Phylogenetische oder idealistische Morphologie?, in: Der Biologe 9, pp. 10-24, 1940.

——, Nochmals: Phylogenetik und Typologie. Entgegnung auf E. Bergdolt: Ueber Formwandlungen, zugleich eine Kritik der Artbildungstheorien, in: Der Biologe 11. pp. 125-129, 1942.

Index